A Practical Guide
to Observational
Astronomy

A Practical Guide to Observational Astronomy

M. Shane Burns

CRC Press
Taylor & Francis Group
Boca Raton London New York

CRC Press is an imprint of the
Taylor & Francis Group, an **informa** business

First edition published 2022
by CRC Press
6000 Broken Sound Parkway NW, Suite 300, Boca Raton, FL 33487-2742

and by CRC Press
2 Park Square, Milton Park, Abingdon, Oxon, OX14 4RN

© 2022 Taylor & Francis Group, LLC

CRC Press is an imprint of Taylor & Francis Group, LLC

Library of Congress Cataloging-in-Publication Data

Names: Burns, M. Shane, author.
Title: A practical guide to observational astronomy / M. Shane Burns.
Description: First edition. | Boca Raton : CRC Press, [2022] | Includes
  bibliographical references and index.
Identifiers: LCCN 2021023320 | ISBN 9780367768638 (hardback) | ISBN
  9781032068022 (paperback) | ISBN 9781003203919 (ebook)
Subjects: LCSH: Astronomy--Observations. | Astronomy--Technique.
Classification: LCC QB145 .B87 2022 | DDC 522--dc23
LC record available at https://lccn.loc.gov/2021023320

ISBN: 978-0-367-76863-8 (hbk)
ISBN: 978-1-032-06802-2 (pbk)
ISBN: 978-1-003-20391-9 (ebk)

DOI: 10.1201/9781003203919

Typeset in Latin Modern font
by KnowledgeWorks Global Ltd.

For Stormy
with love

Contents

Preface

I intend this book to be used as the textbook for an observational astronomy course for upper-level astronomy and physics majors. The book started as a collection of notes for an observational astronomy course I first taught at Harvey Mudd College in 1990. I continued developing and updating the book as I taught similar courses at the U.S. Air Force Academy and Colorado College. These courses have all been observation intensive, leaving little time for traditional lecture-type classroom sessions or extraneous reading. In this text, I've tried to cover only ideas essential to making and analyzing astronomical observations done in an observation-intensive course. A few other modern textbooks on observational astronomy exist, but they tend to be too specialized or too long for a practical observational course. I refer readers to more comprehensive texts to learn more about any subject treated briefly in this text. I especially recommend Fredrick Chromey's book *To Measure the Sky* [7]. It is well written and comprehensive.

Modern astronomy would be impossible without the extensive use of computers, both for control of astronomical instruments and data analysis. Essentially all research-grade telescopes and astronomical detectors are computer controlled. Astronomers need to use software to access and analyze the data they produce. Understanding how to use computers to control equipment and analyze data is as crucial to modern astronomers as a telescope.

Many of the problems in this book require the use of a computer. I haven't specified a particular programming language or software package in the book, but I provide examples and tutorials using the Python programming language on the course's companion website, https://mshaneburns.github.io/ObsAstro/. Python is a free and open-source programming language and runs under Windows, macOS, and Linux operating systems. In a 2015 survey, Momcheva and Tollerud [16] found that Python was the most common programing language used by astronomers. Some of the Python scripts on the site use software developed by the Astropy Project.[1] The Astropy Project is an astronomical community effort to develop a set of packages for astronomy in Python. The Space Telescope Science Institute has begun to phase out most of its old software tools and is moving their functionality to a set of Python packages called Astroconda.[2] Python-based software seems to be the tool of choice for astronomical image processing and data analysis.

Many of the problems in Chapter 5 on image processing require the student to download image files from the companion website. They can display the images using Python, but there are other arguably better alternatives. The Astrobites website[3] contains a summary of the software used by most professional astronomers and includes information about some other options for image display.

The students will need to have Python and several astronomy-related Python packages installed on their computers to complete many of this book's problems. I recommend using the Anaconda Python Distribution.[4] Anaconda is free, open-source, and the basic installation will include all packages students will need. The companion website contains links and instructions for installing Anaconda. The site also has some sample scripts for reading and manipulating image data files. Students can download these scripts from the website and use them as starting points for creating their image processing scripts. All of the software on the site was tested using the Python 3 version of the Anaconda Distribution, so students must download the latest version of Python 3. Of course, the students will also need to be familiar with writing and running simple Python scripts. There are several resources for learning Python listed on the companion website.

I want to express my gratitude to the students and colleagues that have read the book, found errors, and contributed recommendations that improved the text. Many students contributed, but I especially want to thank Don Hoard, Rob Knop, JD Merritt, Michael Leveille, Arnaud Michel, Maddie Lucey, and Zoe Pierrat. I also want to thank my colleagues, Sandy Sandman, Robert Chambers, Bryan Penprase, Jack Wetterer, Natalie Gosnell, and Emily Leiner.

Finally, I wish to thank my life partner, Stormy Burns. Without her help, this book wouldn't exist.

---

[1] https://www.astropy.org
[2] https://astroconda.readthedocs.io/
[3] https://astrobites.org/guides/guide-to-astrophysical-software/
[4] https://www.anaconda.com/distribution/

CHAPTER 1

Astronomical Coordinates and Time

Suppose one night you were looking through your backyard telescope and you discovered what you think is a supernova. If you want to tell others where to look to find this event, you will need to designate the position of the object in the sky. This chapter introduces the standard astronomical coordinates systems that astronomers use to specify the location of objects on the celestial sphere. The **celestial sphere** is an imaginary sphere on which the astronomical objects appear to be located. We can specify any point on the sphere by specifying two angles. Only two angles are needed, but there are a variety of different and surprisingly subtle ways to define those angles.

To specify any coordinate system, we need to choose an origin for the coordinates. In astronomy, this is usually taken to be at the center of the Earth or the Sun. We also need to specify the reference directions from which to define the coordinates. For example, to define a conventional Cartesian coordinate system, we need to specify the directions of the x, y, and z-axes. Astronomical coordinate systems use two angles of a spherical coordinate system. Spherical coordinates are specified by defining a **fundamental plane**, a **fundamental direction**, and the directions of increasing angle. The fundamental plane is a plane which divides the sphere into two hemispheres. The fundamental direction specifies the direction of one of the axes. The traditional definition of spherical polar coordinates defines the x–y plane as the fundamental plane. Points on the surface of a sphere centered on the origin specified by the angles θ and ϕ, where θ is the angle from the $+z$-axis and ϕ is the angle in the

DOI: 10.1201/9781003203919-1

x–y plane shown in Figure 1.1. The angle ϕ increases from the $+x$-axis toward the $+y$-axis.

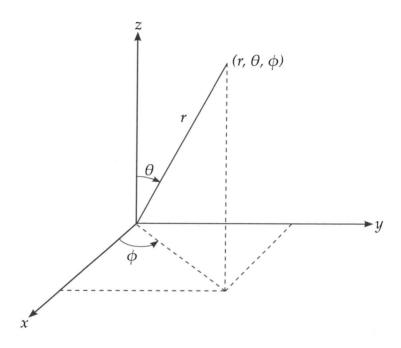

Figure 1.1: Two angles θ and ϕ specify the location of any point on the sphere. This is a conventional right-handed coordinate system, with ϕ increasing from the $+x$-axis toward the $+y$-axis.

1.1 HORIZON COORDINATES

The **horizon** or **altitude-azimuth coordinate system** is the simplest way to describe the position of a celestial object from the surface of the Earth. The origin of horizon coordinates is taken to be the observer. The fundamental plane is a plane tangent to the surface of the Earth at the location of the observer so that the z-axis of a conventional Cartesian coordinate system would point toward the **zenith**. The zenith is defined to be the point on the celestial sphere directly over the observer. The point directly under the observer is called the **nadir**. The fundamental direction for horizon coordinates is due north. The x-axis of a Cartesian coordinates would point in this direction. The position of a star is determined by specifying the **altitude** angle h from the horizon to the star, and an **azimuth** angle A measured along the horizon from north or the x-axis (Figure 1.2). Notice that for conventional horizon coordinates A increases from north to east, which makes this a left-handed coordinate system. The **zenith distance** z is the angle between the z-axis and the

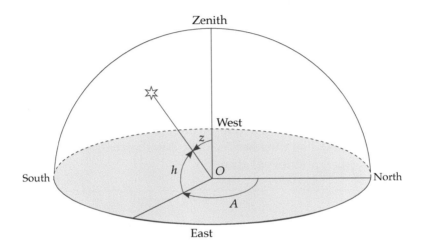

Figure 1.2: The horizon coordinate system.

star so that $z = h - 90°$. The **local meridian** is the arc on the celestial sphere from the point on the horizon that is due south, through the zenith to the point on the horizon that is due north.

Horizon coordinates are useful because they are simple, but have several disadvantages. One is that the origin of the coordinates depends on the location of the observer. This isn't a serious problem for most celestial objects. The stars are so far away that shifting the origin to the center of the Earth wouldn't change the star's altitude and azimuth. A more serious disadvantage is that the directions of the axes depend on the observer's location. The azimuthal coordinate depends on the observer's longitude and the zenith distance depends on the observer's latitude (see Problem 1.8). However, the biggest disadvantage is that celestial objects don't have fixed coordinates. This is due to the Earth's rotation and orbital motion around the Sun.

1.1.1   Diurnal and Annual Motions

As the Earth rotates all the celestial objects rise, transit[1], and set. Their altitude and azimuth are constantly changing. The length of the day is derived from the motion of the Sun around the celestial sphere. The **solar day** is the time between successive transits of the Sun. The time between transits is on an *average* 24 hours, but some days are actually be a little shorter than 24 hours and some a little longer. The difference is due to the fact that the Earth's orbit is slightly elliptical. Section 1.5 discusses this effect in more detail.

If we measured the time between transits of a star rather than the Sun we would find that the time between transits is on average about 23 hours and 56 minutes. This is because as the Earth travels in its orbit around the Sun the position of the Sun shifts with respect to the background stars. The time between the stars' successive transits of the meridian is called a **sidereal day**.

Problem 1.1

Given that the length of the year is 365.242 solar days, show that the length of the sidereal day is approximately 4 minutes shorter than the solar day.

The apparent path of the Sun along the celestial sphere is called the **ecliptic**. As the year progresses the Sun passes through the **zodiacal constellations**. The zodiacal constellations are the thirteen constellations along the ecliptic[2]. The planets' and the Moon's paths along the celestial sphere are very close to the ecliptic.

1.2   EQUATORIAL COORDINATES

Given the fact that the horizon coordinates depend on the position of the observer and that the coordinates of any object in the sky change as the Earth rotates under the celestial sphere, horizon coordinates aren't very useful. A better system would be one that is independent of the celestial motion or the observer's location on Earth. Astronomers use several

---

[1]A **transit** occurs when a celestial body crosses the local meridian.

[2]Traditionally there are twelve zodiacal constellations: Capricornus, Aquarius, Pisces, Aries, Taurus, Gemini, Cancer, Leo, Virgo, Libra, Scorpius, and Sagittarius. However, according to the way astronomers divide the sky into constellations the Sun also passes through the constellation Ophiuchus.

different coordinate systems depending on which is most convenient. Galactic coordinates are used when the location of objects with respect to the plane of the galaxy are important. Ecliptic coordinates are useful for solar system studies, but the most commonly used system is the equatorial coordinate system.

In order to define the two equatorial coordinates consider the location of the star on the celestial sphere in Figure 1.3. The origin of equatorial coordinates is the center of the Earth. The fundamental plane is coincident with Earth's equator. That means the z-axis of a conventional Cartesian coordinate system would be coincident with the Earth's polar axis (see Figure 1.3). The intersection of the Earth's equatorial plane and the celestial sphere defines the **celestial equator**. The intersections of the positive and negative z-axis with the celestial sphere are called the **north celestial pole** (NCP) and **south celestial poles** (SCP). The intersection of the fundamental plane with the celestial sphere is called the **celestial equator** (CE). The two angles that specify the location of a point on the celestial sphere are called the **right ascension** α and the **declination** δ. The declination is the polar angle and is measured from the celestial equator, so $\delta = 90° - \theta$. The right ascension is the angle from the fundamental direction in the fundamental plane as is shown in Figure 1.3. The earth rotates 360° in right ascension in approximately 24 hours. This makes it convenient to specify right ascension in hours, minutes, and seconds of time. With one hour being 15°, 1 minute is 15 arcminutes and 1 second is 15 arcseconds. By convention, the right ascension is a positive angle between 0 and 24 hours.

The fundamental direction—the direction of the positive x-axis—points toward a fixed point on the celestial sphere called the **First Point of Aries** and is usually given the symbol ♈. The First Point of Aries is located at one of the intersections of the ecliptic and the celestial equator. The ecliptic is titled with respect to the celestial equator because the Earth's axis is tilted with respect to the Earth's orbital plane. The time at which the Sun crosses the celestial equator is called an **equinox**. This happens twice a year—once on about September 22nd and once on about March 20th. The position of the Sun at the September equinox defines the First Point of Aries. The northern-most excursion of the Sun occurs around June 21st. This time is called the **June solstice**. The southern-most excursion is called the **December solstice**.[3] Figure 1.4 shows the

---

[3]The traditional names for these events are the autumnal equinox, the winter

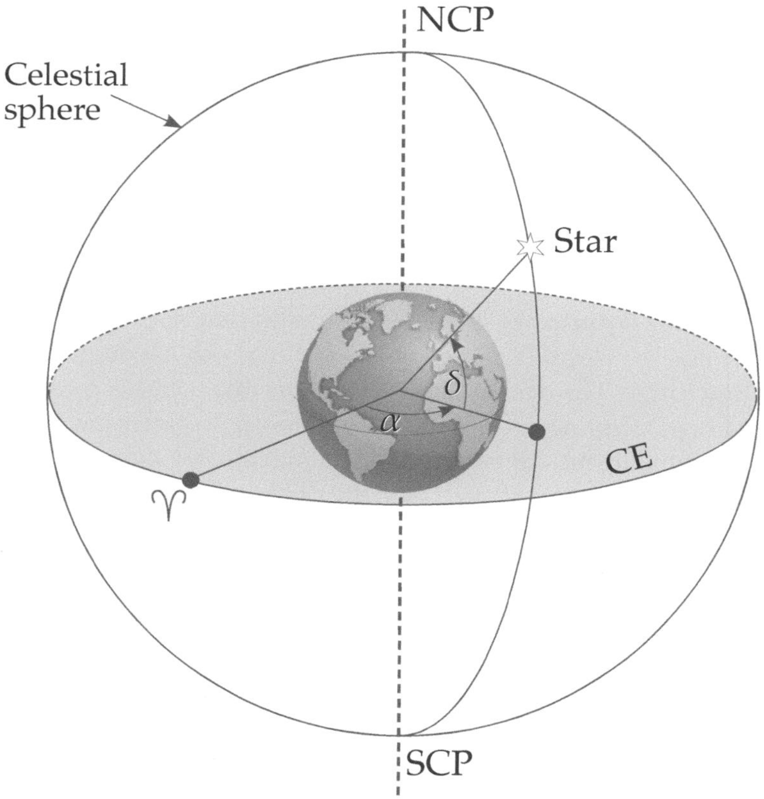

Figure 1.3: The equatorial coordinate system. The z-axis intersects the celestial sphere at the north and south celestial poles, NCP and SCP respectively. The x–y plane intersects the celestial sphere at the celestial equator. The x-axes points toward the first point of Aries (Υ).

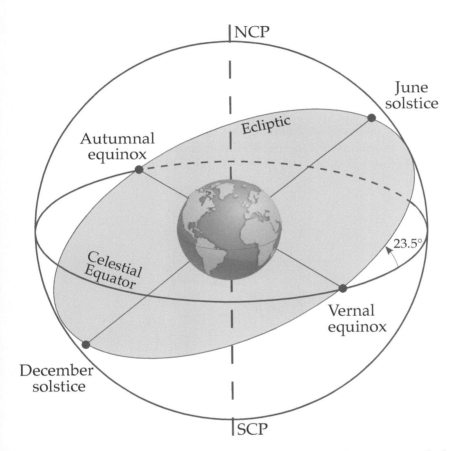

Figure 1.4: The celestial sphere showing the celestial equator and the ecliptic. The locations of the Sun during the solstices and equinoxes are shown.

location of the Sun on the celestial sphere at the times of the equinoxes and solstices.

Problem 1.2
Figure 1.4 is a geocentric view of the path of the Sun around the Earth. Draw an equivalent heliocentric diagram showing the path of the Earth around the Sun. Be sure to show the Earth and the direction of the Earth's axis at the time of the equinoxes and solstices.

1.2.1 Angular Separation in Equatorial Coordinates

You will often find it useful to be able to calculate the angular separation between two points on the celestial sphere from their equatorial coordinates. Figure 1.5 shows two points A and B on the celestial sphere. The

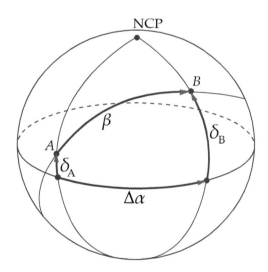

Figure 1.5: Two points, A and B, on the celestial sphere separated by the angle β.

angular separation between the two points is β. The coordinates equatorial coordinates of point A are α_A and δ_A. The coordinates of point B

---

solstice, vernal equinox, and summer solstice, but I prefer terms that avoid the northern hemisphere bias.

are α_B and δ_B. The points A, B, and the NCP defines a spherical triangle on the celestial sphere. Applying the law of cosines for a spherical triangle [equation (A.1)] from Appendix A to this triangle gives

$$\cos \beta = \cos(90° - \delta_A)\cos(90° - \delta_B) \\ + \sin(90° - \delta_A)\sin(90° - \delta_B)\cos(\alpha_B - \alpha_A),$$

or

$$\cos \beta = \sin \delta_A \sin \delta_B + \cos \delta_A \cos \delta_B \cos \Delta\alpha, \qquad (1.1)$$

where $\Delta\alpha \equiv \alpha_B - \alpha_A$ and $\Delta\delta = \delta_B - \delta_A$. For sufficiently small angles $\Delta\delta$ and $\Delta\alpha$ the angle β is

$$\beta^2 = \Delta\delta^2 + \left(\cos^2 \bar{\delta}\right)\Delta\alpha^2, \qquad (1.2)$$

where $\bar{\delta} = \frac{\delta_A + \delta_B}{2}$.

Problem 1.3
Starting with equation (1.1) prove equation (1.2) for small angles $\Delta\delta$ and $\Delta\alpha$.

1.2.2 Relating Equatorial to Horizon Coordinates

Figure 1.6 shows the celestial sphere with the reference points for both horizon and equatorial coordinates. The actual horizon with the cardinal points (North, South, East, and West) are shown in the horizontal plane. The zenith is the point directly above the observer. The celestial equator lies in a plane tilted with respect to the horizon. The angle along the meridian between the horizon plane and the celestial equator is equal to 90° minus the latitude of the observer. The latitude is also the angle between the NCP and the zenith. The figure also shows the intersection of two planes. One plane contains the north celestial pole, the zenith and the south celestial pole. The intersection of this plane with the celestial sphere is the meridian. Another plane is defined by the north celestial pole, a star (not on the meridian) and the south celestial pole. The **hour angle** (HA) is the angle, measured along the celestial equatorial, between the plane containing the meridian and the plane containing the star. From Figure 1.6, it is easy to see that

$$HA = \alpha_M - \alpha, \qquad (1.3)$$

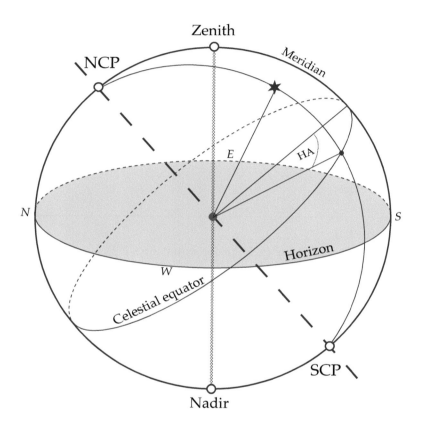

Figure 1.6: Diagram of the celestial sphere showing the meridian plane and the object plane. The meridian plane is the plane containing the north celestial pole (NCP) the south celestial pole, SCP and the zenith. The object plane contains the object and the north and south celestial poles. The angle, on the celestial plane, between the meridian plane and the object plane is the hour angle.

where α_M is the right ascension of the meridian. The right ascension of the meridian is equal to the **local sidereal time** or LST. Sidereal time is a time system based on the length of the sidereal day, but it is best to think of the LST as simply the right ascension of the meridian

$$LST = \alpha_M. \qquad (1.4)$$

Like the right ascension, the hour angle is typically stated in hours, minutes, and seconds. Unlike the convention for right ascension, hour angle ranges from -12 hours to $+12$ hours. All points on the celestial sphere east of the meridian have negative HA while those in the west have positive HA.

In order to point the telescope at an object, one needs to know its equatorial coordinates and α_M. Fortunately, most research telescopes are computer controlled so the computer automatically calculates LST for you. However, it is still useful to know how to approximate α_M so that one can determine what right ascensions are visible at a particular time on a given night.

It is easy to derive an approximation for α_M from the definition of the first point of Aries and the September equinox. By definition, the Sun has a right ascension of zero at the time of the March equinox. Since the Sun transits the meridian at about noon local time, $\alpha_M \approx 0\,\mathrm{hr}$ at noon local time on the first day of spring. One hour later, $\alpha_M \approx 1\,\mathrm{hr}$ etc. At midnight $\alpha_M \approx 12$ hrs, so the right ascension visible at midnight on the night of the March equinox ranges from $\alpha \approx 6\,\mathrm{hrs}$ to 18 hrs. Note that this is actually only true for objects on the celestial equator. In the northern hemisphere a larger range of right ascensions is visible for objects north of the celestial equator and a smaller range for those to the south.

Six months later, on the day of the September equinox, the Sun crosses the celestial equator and has a right ascension, $\alpha \approx 12\,\mathrm{hrs}$ so $\alpha_M \approx 12$ hrs at noon. This implies that at *midnight* local time on about September 22nd $\alpha_M \approx 0$ hrs. It takes one year or approximately 365 days for the sun to return to this position so α_M at midnight can be computed on any night of the year by multiplying the fraction of the year since the September equinox by 24 hrs.

$$\alpha_M(\text{at midnight}) \approx \frac{n}{365}\ 24\ \mathrm{hrs}, \qquad (1.5)$$

where n is the number of days since the September 22nd. This is only an approximation! To calculate the exact value we need to know the exact time the Sun crosses the celestial equator and the longitude of our telescope. However, the above equation is good to a few percent and is sufficiently accurate for planning most observations.

Once we have computed α_M at midnight, it is easy to determining α_M for any time during the night. For example, if $\alpha_M = 13$ hrs at midnight then at 9:00 PM, $\alpha_M \approx 13\text{hrs} - 3\text{hrs} = 10$ hrs.

Problem 1.4

Suppose we wish to use the Keck Telescope on Mauna Kea to observe the galaxy M51. The coordinates of M51 are $\alpha \approx 13^{\text{h}}30^{\text{m}}$ and $\delta \approx 47°13'$. When is the best time to observe? That is, on about what night of the year will M51 transit at midnight?

1.2.2.1 Transformations for Equatorial and Horizon Coordinates

Suppose we knew the equatorial coordinates of some celestial object and we wanted to know the horizon coordinates. We would need to not only know α and δ, but also the local sidereal time, α_M and latitude, ϕ, of the observatory. Given all this information we can use the law of cosines and law of sines for spherical triangles to determine the altitude, h, and azimuth, A.

$$\sin h = \sin \delta \sin \phi + \cos \delta \cos(HA) \cos \phi \qquad (1.6)$$
$$\cos h \sin A = \cos \delta \sin(HA), \qquad (1.7)$$

where $HA = \alpha_M - \alpha$.

Problem 1.5

What is the azimuth of the Sun at sunset on the December solstice as viewed from Kitt Peak National Observatory (latitude 31.9583° N, and longitude 111.5967° W)?

Notice that these transformations imply the $h > 0$ for $\delta > \phi$ for all HA. That means that these objects never set. An object with a declination such that its altitude is never negative for a given latitude is said to be **circumpolar** and will be visible anytime during the year. Objects with $\delta < -\phi$ have $h < 0$ for all HA. These objects never rise and are hence are never visible from that latitude. Figure 1.7 shows the locations of circumpolar objects and objects that never rise on the celestial sphere.

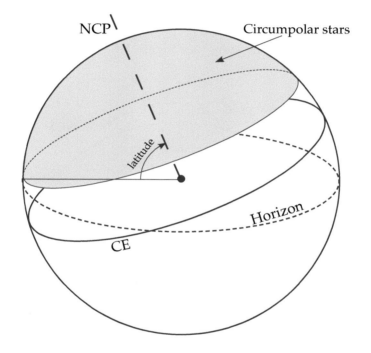

Figure 1.7: Celestial sphere showing the locations of objects that are circumpolar.

1.2.3 Precession and Nutation

The fundamental plane and the fundamental direction for equatorial coordinates are tied to the axis of rotation of the Earth. If there were no torques on the Earth, its angular momentum would be conserved and the axis of rotation would be fixed. However, the gravitational forces of the Moon, Sun, and planets on the equatorial bulge of the Earth produce a torque that changes the direction of its axis of rotation. The change in the axis of rotation is broken down into two components called precession and nutation. **Precession** is a slow movement of the Earth's axis of rotation around an axis perpendicular to the ecliptic, which is fixed with respect to the stars (see Figure 1.8). It takes approximately

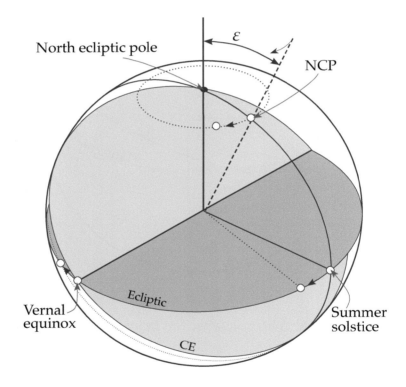

Figure 1.8: Precession of the equatorial coordinate system. The ecliptic poles remains fixed with respect to the stars, but the north and south celestial poles as well as the position of the First Point of Aries changes as the Earth's axis precesses.

26,000 years for the axis of rotation to make one complete cycle around the ecliptic pole. The angle between the ecliptic pole and the celestial pole is called the **obliquity** and is approximately 23.4° for the Earth. As a consequence of this motion, the first point of Aries also precesses with a period of 26,000 years. Both the fundamental plane and the fundamental direction are changing slowly with time. Currently the north celestial pole is close to the bright star Polaris, but in about 14,000 years the star Vega will be near the NCP making Vega the pole star. Currently the First Point of Aries is in the constellation Pisces, but in 1000 BCE it was in the constellation Aries.

Problem 1.6

Estimate how much and it what direction the First Point of Aries precesses each year.

In addition to precession, there is a relatively short term small oscillation around the precessional path called **nutation**. Nutation causes the positions of celestial objects to oscillate sinusoidally by almost 20″ with a period of 19 years. Nutation occurs because the torque exerted by the Sun, Moon, and planets isn't constant in time. The nutation is effected not just by the gravitational forces, but also by changes in the moment of inertia of the Earth. This makes the detailed calculations of the nutation difficult, but fortunately these effects are small compared to the effects of precession.

The effects of precession and nutation causes the equatorial coordinates of celestial objects to change over time. This means that it isn't enough to just specify the coordinates, we must also specify the date or **epoch** for the coordinates. All catalogs that specify the coordinates of astronomical objects also specify an epoch. One of the most common epochs used today is the so called J2000 epoch. J2000 refers to 12 h GMT on January 1st, 2000.[4]

Given the complicated motion of the celestial pole with respect to the stars, calculating the *exact* coordinates of an object for today when given the position at a different epoch is a difficult task. However, most of the change in coordinates is due to precession and a good approximation is

---

[4]GMT stands for Greenwich Mean Time and the "J" in J2000 refers to the Julian date. See Section 1.5 on time for a complete description.

given by the equations

$$\alpha = \alpha_0 + (m + n \sin \alpha_0 \tan \delta_0) \, N \tag{1.8}$$
$$\delta = \delta_0 + (n' \cos \alpha_0) \, N, \tag{1.9}$$

where α and δ are the current-day coordinates. The constants m, n, and n' are given in Table 1.1. The coordinates α_0 and δ_0 are the coordinates at the epoch listed in Table 1.1 and N is the number of years since the reference epoch.

Table 1.1: Precessional constants from Duffett-Smith and Zwart [9].

Epoch	m (seconds/year)	n (seconds/year)	n' (arcsec/year)
1900.0	3.07234	1.33645	20.0468
1950.0	3.07327	1.33617	20.0426
2000.0	3.07420	1.33589	20.0383
2050.0	3.07513	1.33560	20.0340

Problem 1.7
The star Sirius has equatorial coordinates $\alpha(\text{J2000}) = 6^{\text{h}} \, 45^{\text{m}} \, 8.9^{\text{s}}$ and $\delta(\text{J2000}) = -16°42'58''$. What are the coordinates on June 5, 2013?

1.3 TELESCOPE MOUNTS

The functions of a telescope mount are to hold the optical elements (lenses and mirrors) in place and to point and track an object of interest. For large telescopes holding the optics in position requires some very sophisticated engineering. For example, the Keck 10-meter telescope has a primary mirror made-up of 36 individual hexagonal segments. The positions of all 36 sections are actively monitored and adjusted in real time. Fortunately, as an astronomer you can remain blissfully ignorant of all this engineering. However, you do need to be aware of the advantages and disadvantages of the two most common types of telescope mounts. These two types of mounts are closely tied to the two coordinate systems we've discussed so far.

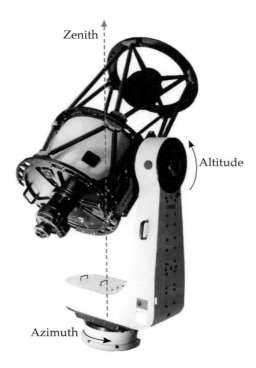

Figure 1.9: Image of telescope on an alt-az mount. The two angles that specify where the telescope is pointing are the altitude and azimuth. (Photo courtesy of PlaneWave Instruments.)

1.3.1 Altitude-Azimuth Mount

The simplest telescope mount is an altitude-azimuth or **alt-az mount** (see Figure 1.9). This type of mount is naturally related to the horizon coordinates. The telescope rotates around a vertical axis to set the azimuth and around a horizontal axis to set the altitude. The main advantage of this system is its simple structural design. It's also more compact than an equatorial mounted telescope, discussed below, and this means it is cheaper to build. The disadvantages are that it must be driven at a non-uniform rate around both axes to track an astronomical object across the sky and the field of view rotates as the telescope tracks. Never-the-less its simplicity and, hence its low cost, make it an attractive design for large telescopes. Most large radio telescopes as well as the Keck ten-meter telescope use this design.

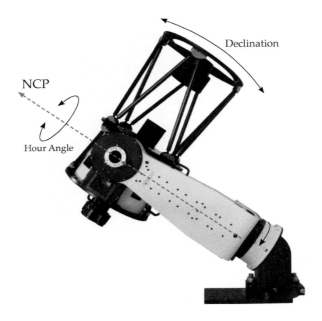

Figure 1.10: Side view of an equatorial mounted telescope. The two angles that specify where the telescope is pointing are the hour angle and declination. (Photo courtesy of PlaneWave Instruments.)

1.3.2 Equatorial Mount

The equatorial mount is the most common design for astronomical telescopes. In this design one of the two axes, the polar axis, is aligned parallel to the Earth's rotational axis, and the other, the declination axis, is perpendicular to the polar axis (see Figure 1.10). This design overcomes the major disadvantages of the alt-az design. In order to track an object, the telescope rotates about the polar axis at a constant rate equal to the Earth's rotation rate, but in the opposite direction; this also means that the field of view of an equatorial mounted telescope doesn't rotate. A final advantage is that the angular measures about the two axes give the hour angle and declination directly.

1.4 OTHER COORDINATE SYSTEMS

1.4.1 Galactic Coordinates

When studying the galaxy, it makes sense to specify the positions of objects with respect to the center and the plane of the galaxy. The origin of galactic coordinates is the Sun. The fundamental plane coincides with the galactic plane and the fundamental direction points toward the center of the galaxy in the constellation Sagittarius. The galactic latitude, b, is the angle for the object above the galactic plane. The galactic longitude, ℓ, is the angle in the galactic plane from the center of the galaxy. By convention, both angles are measured in decimal degrees with $-90° \leq b \leq 90°$ and $0° \leq \ell \leq 360°$. Figure 1.11 shows an "image" of the Milky Way galaxy and shows how galactic coordinates of a star are defined.

The transformation from equatorial to galactic coordinates for J2000 is [8]

$$\sin b = \sin \delta \cos 62.87° - \cos \delta \sin(\alpha - 282.86°) \sin 62.87° \tag{1.10}$$

$$\cos b \cos(\ell - 32.93°) = \cos \delta \cos(\alpha - 282.86°). \tag{1.11}$$

1.4.2 Ecliptic Coordinates

It is convenient to use ecliptic coordinates when specifying the positions of solar system objects—planets, minor planets, comets, etc. Ecliptic coordinates have the Earth as the origin, but the fundamental plane is coincident with the ecliptic plane. Like equatorial coordinates the fundamental direction is toward the First Point in Aries. The ecliptic latitude, β, is the angle from the fundamental to the object. The ecliptic longitude, λ, is the angle in the ecliptic plane from the First point of Aries. It is a right-handed coordinate system and by convention, both angles are measured in decimal degrees with $-90° < \beta < 90°$ and $0° < \lambda < 360°$. Ecliptic coordinates are related to equatorial coordinates by a simple rotation around the fundamental direction by an angle equal to the obliquity, ϵ (see Figure 1.12). The transformation from equatorial to ecliptic coordinates is

$$\tan \lambda = (\sin \alpha \cos \epsilon + \tan \delta \sin \epsilon)/ \cos \alpha \tag{1.12}$$

$$\sin \beta = \sin \delta \cos \epsilon - \cos \delta \sin \epsilon \sin \alpha. \tag{1.13}$$

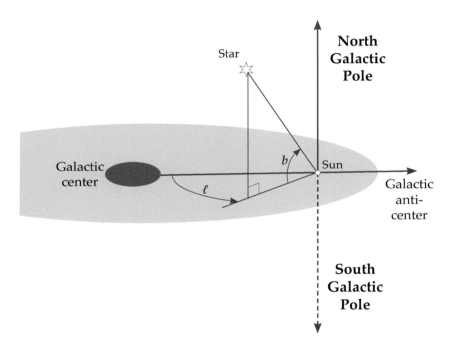

Figure 1.11: The galactic coordinate system. Galactic coordinates have their origin at the Sun. The galactic latitude, b, is measured from the galactic plane and the galactic longitude ℓ is angle in the galactic plane measured from the galactic center.

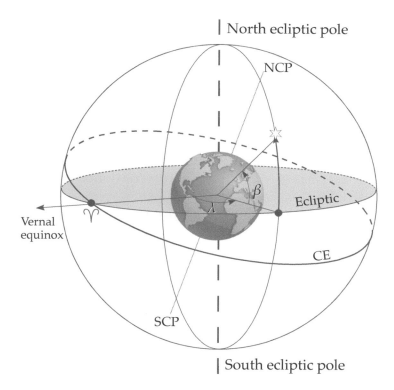

Figure 1.12: Ecliptic coordinates are like equatorial coordinates rotated by an angle equal to the obliquity around the axis pointing toward the First Point of Aries.

1.5 TIME IN ASTRONOMY

All of our measures of time were originally derived from intervals between celestial events. The length of the day was taken as the time between two successive transits of the Sun. The day is subdivided into 24 hours; each hour is divided into 60 minutes and each minute into 60 seconds. A second could then be defined as 1/86,400 of a solar day. The problem with this definition is that not all solar days are the same length. This is primarily due to the ellipticity of the Earth's orbit. The problem is avoided by defining the **mean solar day** as the average length of the solar day over an entire year. Even this doesn't completely solve the problem because the Earth's rate of rotation varies slightly and is slowing overall. Today, the **International System of Units** (abbreviated **SI**) defines the **SI second** to be the length of time equal to 9,192,631,770 periods of the radiation originating from the transition between two hyperfine energy levels of the cesium-133 atom. This definition was chosen so that it is equal to 1/86,400 of the mean solar day in 1900.

The SI second is the basic unit of **International Atomic Time (TAI)**. TAI is from the French name *temps atomique international*. Having a time system that is based on atomic clocks means the system is accurate to less than a fraction of a second over a million years. The problem with this system is that it gets out of sync with astronomical time because of variations in the Earth's rotation rate. TAI is currently about 35 seconds out of sync with the position of the Sun. **Coordinated Universal Time** or **UTC** is a time standard that is based on SI seconds, but is kept in sync with the Sun by occasional additions of a leap second.

1.5.1 Solar Time

There are two reasons the Sun is a poor time keeper. One is that the Earth's orbit is slightly elliptical—it moves faster at perihelion in January and more slowly at aphelion in June. The second effect is due to the tilt of the Earth's axis with respect to its orbital plane. From the Earth's point of view, this means that the ecliptic is tilted with respect to the equatorial plane. The Sun's apparent diurnal motion depends on the position of the Sun on the ecliptic projected onto the celestial equator. The net effect is for the solar day to be slightly longer than 24 hours in May and November and slightly shorter than 24 SI hours in February and July. The **apparent solar time** or **true solar time** is based on the actual position of the Sun on the sky. It is equal to the HA of the Sun plus 12 hours or

$$T_{AS} = HA_{\odot} + 12\,\text{hr},$$

where T_{AS} is the apparent solar time and, $HA_{\odot}$ is the hour angle of the Sun. At noon local time the apparent solar time should be very close to 12 hr.

The **mean solar time** is defined by the position of a fictitious Sun that moves across the sky at a rate equal to the mean rate of the real Sun. At noon mean solar time, the fictitious mean Sun transits the meridian. The **solar day** is defined as the time interval between two successive transits of the fictitious mean sun and is *very* close to 24 SI hours long. The **equation of time** (EOT) is the difference between the mean solar time and the apparent solar time

$$EOT = T_{MS} - T_{AS},$$

where T_{MS} is the mean solar time. You can see in Figure 1.13 that the apparent Sun can be as much as 15 minutes ahead of or behind the fictitious mean Sun. The time origin for UTC is chosen to match the

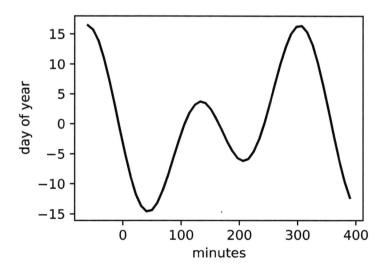

Figure 1.13: Equation of time showing by how much the actual position of the Sun differs from the mean Sun.

mean solar time at 0° longitude or the Royal Observatory in Greenwich

England. When UTC becomes a second out of sync with the mean solar time at Greenwich, which is called **UT1**[5], a leap second is added to UTC so it always agrees with UT1 to within 0.9 seconds.

Our system of standard time zones are referenced to UTC. Each zone consists of a band about 15° of longitude wide. Figure 1.14 shows the regions for each of the standard zones. Some parts of the world also institute daylight savings time. During daylight savings time the offset between UTC and the time in the zone is one hour less. For example, Mountain Standard Time (MST) is 7 hr behind UTC, but Mountain Daylight Time (MDT) is 6 hr behind so at 22:30 UTC it is 15:30 MST, but 16:30 MDT.

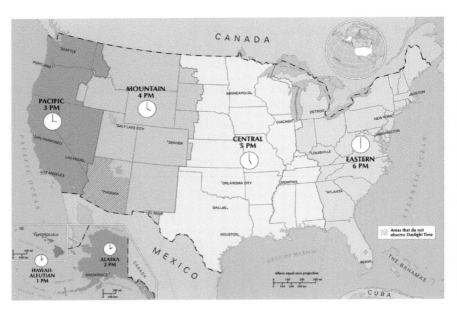

Figure 1.14: United States time zones. (Source: Wikimedia Commons)

1.5.2   Julian Date

In 1054, Chinese astronomers reported the discovery of a "guest star" in the constellation Taurus. Today with modern telescopes we see a supernova remnant in exactly the same location. We know it as the Crab

---

[5]UT1 is the successor to Greenwich Mean Time (GMT). When the small distinction between UTC and UT1 isn't important, they are both often referred to as Universal Time or UT.

Nebula. Apparently the guest star was the original supernova. Suppose as an astronomer you wished to *accurately* determine the interval of time between the supernova event and today. If you were given the calendar dates you would have to first count the number of years between the events and multiply by 365, then add the appropriate number of leap days, count the number of days in the any additional months and finally count any additional hours. As you can imagine this is a very tedious process. Such calculations arise frequently in astronomy. In 1582, the system of **Julian Day Numbers** was introduced to avoid this sort of calculation. In this system, the time of an astronomical event is specified as the decimal number of days since noon UTC on January 1, 4713 BC. This is called the **Julian Day** or **JD** of the event. Julian Day numbers for each day of the year at 0 UT are tabulated in the *Astronomical Almanac* [22]. For example, the Julian date for 0 UT August 15, 1997 is 2,450,675.5. *The Astronomical Almanac* is published annually by the United States Naval Observatory and Her Majesty's Nautical Almanac Office. It contains ephemera for solar system objects, catalogs of celestial objects as well as information about timekeeping and a list of the world's major observatories. There is an online version at http://asa.hmnao.com.

The Julian Day System was introduced by Joseph Justus Scaliger. There is no direct connection between the Julian Day System and the Julian Calendar used by the Romans. The starting date of January 1, 4713 BC also has no astronomical or historical significance. It was chosen by Scaliger to simplify the calculation of JD for some historic astronomical events.

A more convenient, but related, system is the **Modified Julian Day (MJD)** system. The **MJD** is the JD minus 2,400,000.5. At 0 UT on January 1, 1997 the MJD was 50,448.0.

It is common for astronomers to use the Julian date to specify the epoch for the equatorial coordinates of a celestial object. For example, the Julian epoch 2000.0 is denoted J2000.0 and refers to January 1.5, 2000. The Julian date for January 1.5, 2000 is JD 2,451,545.0 or is the MJD 51,544.5.

1.5.3 Sidereal Time

The apparent solar time is based on the actual position of the Sun. A solar day is the length of time between successive transits of the Sun. A **sidereal day** is the time between successive transits of the First Point

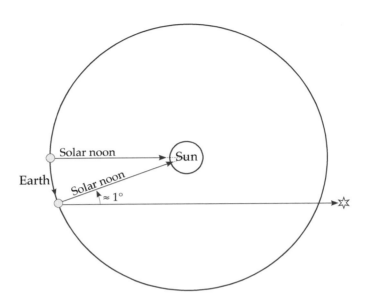

Figure 1.15: Sidereal and solar days. As the Earth rotates once with respect to the fixed stars, it also moves through a certain distance in its orbit. To complete a mean solar day, it must rotate approximately 1° more to bring the Sun back to the meridian.

of Aries. Sidereal time is based on the diurnal motion of the celestial sphere rather than the Sun.

The length of the mean solar day is 24 hours. This is not, however, the period of rotation of the Earth with respect to the celestial sphere. Because the earth is orbiting the Sun, the apparent position of the Sun on the celestial sphere changes from day to day. The rotation period of the Earth is defined by the time it takes a fixed point on the celestial sphere to make one cycle around the Earth. The period of the Earth's rotation is therefore one sidereal day.

A sidereal day is approximately four minutes shorter than the mean solar day. It is easy to derive this time difference. Figure 1.15 shows how the Earth has moved along its orbital path during one period of Earth rotation. The Earth must rotate slightly more than 360° before the Sun transits the local meridian. Since the Sun makes one complete cycle of 360° around the celestial sphere in one year, it must move 360°/365 days

$\approx 1°$ per day with respect to the celestial sphere. The length of the solar day, 24 hours, is actually the amount of time it takes the Earth to rotate $361°$ so that the Sun is on the local meridian. The time, ΔT, needed for the Earth to rotate that extra degree is

$$\Delta T = \left(\frac{24 \text{ hrs}}{361}\right) \approx 0.066 \text{ hrs} \approx 4 \text{ minutes.}$$

A more detailed calculation of the length of the sidereal day gives 23 h 56 m 04 s or 23.9345 hours. The exact ratio of the solar day to the sidereal day is 1.00273790931.

The **local sidereal time** (LST) is defined to be the hour angle of the First Point of Aries. The First Point of Aries defines the fundamental direction so the right ascension of the First Point of Aries is zero, $\alpha_\Upsilon = 0$. Substituting $\alpha_\Upsilon = 0$ into the definition of HA gives $LST = \alpha_M$ [equation (1.4)]. The right ascension of the meridian is equal to the local sidereal time. Calculating the LST accurately requires some care, but it is easy to compute an approximate value as was done in Section 1.2.2.

SUPPLEMENTARY PROBLEMS

Problem 1.8 Suppose an star is seen at the zenith as viewed from Greenwich England.

(a) What is the altitude and azimuth of the star viewed by an observer on the Earth's equator directly south of Greenwich?

(b) What is the altitude and azimuth of the star as viewed by an observer directly East of Greenwich at a longitude of 20° E?

(c) What is the altitude and azimuth of the star as viewed by an observer in Madrid at 40.4000° N and 3.6833° W?

Problem 1.9 Estimate the LST on January 31, 1954 at 10:00 PM.

Problem 1.10 Compute as accurately as possible the azimuth of the Sun at sunset on the day of the December solstice

(a) at Greenwich England.

(b) at Colorado Springs, Colorado.

CHAPTER 2

Optics and Telescopes

The essence of observational astronomy is the measurement of electro-magnetic radiation—light. To learn about the universe astronomers use the whole electromagnetic spectrum from long-wavelength radio waves to high-energy γ-rays. We can only measure a few things about electromagnetic radiation. We can measure the intensity. Astronomers call this **photometry**. Under the heading of photometry we might also include imaging since this is just measuring intensity of the radiation from different parts of the object. We can take a spectrum of the object (**spectroscopy**). Finally, we can measure the polarization of the radiation (**polarimetry**).

Each of these techniques is used to learn different things about the object of interest. For example, we can use photometry to learn about the energy output of a star and obtain a spectrum to determine its temperature. Polarimetry can sometimes be used to determine the emission mechanism since some mechanisms, synchrotron emission for example, produce polarized light.

All of these measurement techniques require a telescope to gather light and focus it on a detector. In this text we will primarily discuss optical telescopes. However, most of the design principles apply to all kinds of telescopes from the largest radio telescopes to space-based x-ray telescopes.

2.1 GEOMETRIC OPTICS

Light is an electromagnetic wave with wavelengths in the range from about 350 to 750 nm. Geometric optics treats light as consisting of narrow beams or rays. The rays are perpendicular to the light's wavefronts.

DOI: 10.1201/9781003203919-2

Table 2.1: Indices of refraction. Fused silica and florite are both used in high quality optics. Data from Allen [8].

Material	n (500 nm)	n (700 nm)
Air (STP)	1.000294	1.000290
Water	1.336	1.330
Fused silica	1.463	1.455
Florite (CaF$_2$)	1.437	1.432

Geometric optics ignores many of the wave and particle properties of light, but works well to describe how lenses and mirrors affect light.

We can determine the paths of light rays by using **Fermat's Principle**. It states that the path that light takes through any medium from one point to another is the one that minimizes the travel time between the two points. The time interval depends on the speed of light. Light travels at speed c in a vacuum, but at a speed less than c in any transparent material. The **refractive index**, n is the ratio of the speed of light in a material to the speed of light in a vacuum

$$n = \left(\frac{c}{v}\right), \tag{2.1}$$

where v is the speed of light in the material and may depend on wavelength. Table 2.1 lists the refractive indices of a few materials.

The travel time, Δt, between two points P_1 and P_2 is given by the path integral

$$\Delta t = \int_C \frac{ds}{v} = \frac{1}{c} \int_C n \, ds \tag{2.2}$$

where the path C intersects points P_1 and P_2. If we define the **optical path length** as

$$S = c \, \Delta t = \int_C n \, ds. \tag{2.3}$$

Minimizing the travel time is equivalent to minimizing the optical path length.

The mathematical problem is to find the curve C that minimizes the integral in equation (2.3). Calculus of variations is the branch of mathematics that studies how to solve such problems in general, but it's easy to see the solution for the simple case where the index of refraction is uniform throughout the medium. In this case the integral for S is just

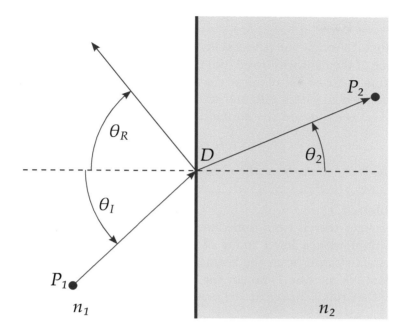

Figure 2.1: Reflection and refraction at the interface between two materials of differing indices of refraction. The index of refraction $n_1 < n_2$ so the ray is bent toward the normal as it passed into the higher index of refraction material.

n times the path length and we know the shortest path between two points is a straight line.

Using Fermat's Principle seems like a lot of work to get something we already knew—that light rays travel in straight lines in a uniform medium—but it is very powerful and can tell us how light travels in *any* medium. In particular you can use it to determine what happens to light when it is incident on the interface between two materials of different indices of refraction. Figure 2.1 shows light rays from point P_1 striking the surface between the two materials at point D. The angle that the light makes with respect to the normal to the surface is called the **angle of incidence** and is labeled θ_I. Some of the light is reflected from the surface. The **law of reflection** is easy to derive from Fermat's Principle

(Problem 2.12) and states that

$$\theta_I = \theta_R \tag{2.4}$$

where θ_R is the angle the reflected light makes with respect to the normal.

Some of the light will travel into the material with index of refraction n_2. Fermat's Principle requires that the light ray change direction or **refract** when it passes into the second material. If $n_2 > n_1$, then the ray will be bent *toward* the normal to the surface so that

$$n_1 \sin \theta_I = n_2 \sin \theta_2 \tag{2.5}$$

where θ_2 is the angle from the normal (Problem 2.13).

Problem 2.1

When light travels from a medium of higher index of refraction to a medium with a lower index of refraction, the incident ray will be refracted *away* from the normal (see Figure 2.2). If the angle of incidence is increased enough the angle of refraction becomes 90°. Show that this happens at the **critical angle**

$$\theta_c = \sin^{-1}\left(\frac{n_2}{n_1}\right). \tag{2.6}$$

What actually happens in cases where $\theta_I \geq \theta_c$ is that *all* of the light is reflected. This is called **total internal reflection**.

All astronomical optics use refraction, reflection, or a combination of the two. In the next few pages we will explore some of these applications.

2.1.1 Prisms

Prisms are optical components with two or more plane surfaces and are used to either disperse or redirect light. You are probably most familiar with how prisms are used to disperse light into its spectral components. Figure 2.3 shows a prism with a ray of light passing through it. The angle α is the angle at the apex of the prism, θ is the angle of incidence at the first surface that the ray encounters, and ϕ is the angle through which the ray is bent with respect to the direction of the incident ray. The ray is refracted twice: once as is passes into the prism and once as it exits the prism. Applying Snell's law twice, once at each surface, gives

$$\sin(\alpha + \phi - \theta) = \sin \alpha \sqrt{n^2 - \sin^2 \theta} - \cos \alpha \sin \theta, \tag{2.7}$$

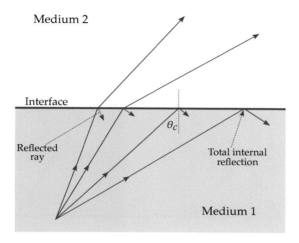

Figure 2.2: The incident ray is traveling from a medium 1 into medium 2 with $n_1 > n_2$. For incident angles greater than the critical angle, $\theta_I \geq \theta_c$, all of the light is reflected back into medium 1.

where n is the index of refraction of the prism. The index of refraction depends on wavelength so given an incident ray of white light, each wavelength will be refracted through a different angle ϕ.

Problem 2.2

Suppose white light is incident on the prism shown in Figure 2.3 at an angle $\theta = 35°$ and that the apex angle $\alpha = 60°$. What are the refracted angles for green light ($\lambda = 500$ nm) and red light ($\lambda = 700$ nm)? Assume the prism is made of fused silica.

Prisms are also used to change the directions of rays or invert images. Prisms used in this way often take advantage of total internal reflection and if designed properly don't produce any dispersion. Figure 2.4 shows how a right angle prism redirects rays by 90° and inverts the image.

2.1.2 Lenses

An **optical lens** uses refraction to cause light rays to converge or diverge. There are a huge number of different kinds of and uses for lenses. We will

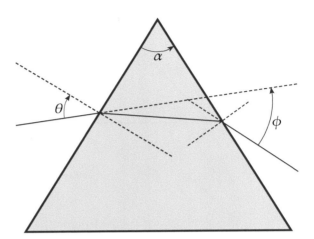

Figure 2.3: A prism with a triangular cross section and index of refraction n.

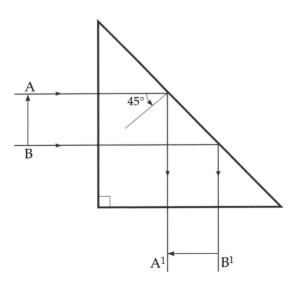

Figure 2.4: A reflecting right-angle prism redirects the rays by 90° and inverts the image.

only explore how lenses are used to form images. For simple **thin lenses**, light rays from an object are refracted in such a way that an image is formed[1]. The **optical axis** of a thin lens is a straight line passing through the geometrical center of a lens and joins the two centers of curvature of its two surfaces. The distance to the image on the optical axis from the center of the lens satisfies the **thin lens equation**,

$$\frac{1}{i} + \frac{1}{o} = \frac{1}{f},\tag{2.8}$$

where o is the distance from object to the lens, i is the distance from the lens to the image, and f is the focal length. The **focal length** is the distance from the lens to the focal point. The **focal point** is the point to which rays that are initially parallel to the axis of the lens are converged or from which they appear to diverge. Figure 2.5 shows the

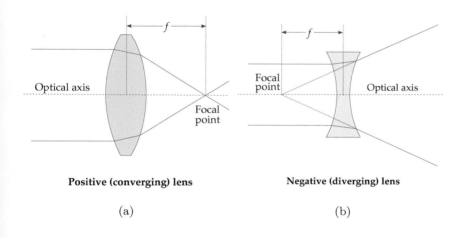

Positive (converging) lens   Negative (diverging) lens

(a)   (b)

Figure 2.5: Focal points for converging and diverging thin lenses. By convention, a converging lens has positive focal lengths, $f > 0$, while diverging lenses have $f < 0$.

locations of the focal point for both a converging and a diverging lens.

A **converging lens** causes rays from a object to be refracted toward the axis of the lens (see Figure 2.6). By convention, the object distance, o,

---

[1]A thin lens is thin in the sense that its thickness is small compared to the image and object distances.

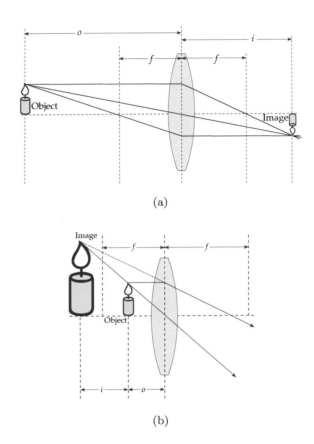

Figure 2.6: Images formed by converging lens. Figure (a) shows the image position when object is outside the focal length of the lens ($o > f$). Figure (b) shows the image position for $o < f$.

is always positive in equation (2.8), as is the focal length of a converging lens. If the image distance is positive the images is on the side of the lens *opposite* the object and if negative the image is on the same side of the lens as the object. Converging lenses refract rays towards the symmetry axis of the lens so by Snell's law one or both of the surfaces of a converging lens must be convex.

Problem 2.3

Use the thin lens equation (2.8) to show that for a converging lens, the image is always on the side opposite of the object for $o > f$, and on the same side as the object for $o < f$.

Images that are formed when rays converge to the image so that the rays actually go through the image are called **real images**. Images from which rays diverge, but don't actually go through the image are called **virtual images**. Figure 2.6(a) shows that the image formed by a converging lens when $o > f$ is a real image, while Figure 2.6(b) shows that a virtual image is formed by a converging lens when $o < f$.

For **diverging lenses**, one or both surfaces are concave so that rays are refracted *away* from the axis of the lens (see Figure 2.7). The focal lengths of diverging lenses are negative by convention. Image distances for diverging lenses are always negative so the images formed by diverging lenses are always on the same side of the lens as the object and are always virtual images.

Problem 2.4

Use the thin lens equation (2.8) to show that the image distance is always negative for diverging lenses.

2.1.2.1 Simple Refracting Telescopes

Galileo was the first to use a telescope for astronomy and document his observations. It used one converging lens and one diverging lens. Figure 2.8(a) shows the optical design for Galileo's telescope. The centers of the two lenses define the optical axis. In the figure, light entered from the top through the converging lens. By itself, the converging lens would bring the light to a focus at point F, but the diverging lens refracts the rays away from the optical axis. The converging lens at the top of the

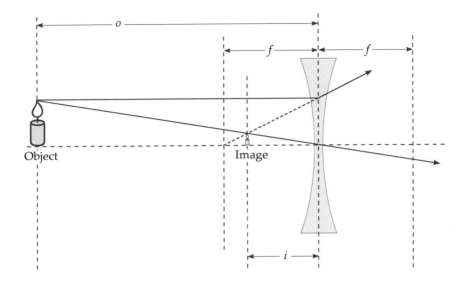

Figure 2.7: Ray diagram for a diverging lens. Note that only virtual images are formed by diverging lenses.

figure, closest to the object being studied, is called the **objective lens**. For astronomical objects, the objects are essentially located at infinity. If the diverging lens wasn't present the converging lens would focus light to a point at F. Point F is located at focal point of the objective lens at distance f_o from it. The lens at the right is called the **eyepiece lens** and causes the light to diverge again into parallel rays so they may be focused by the eye. The net effect is to compress all of the rays passing through the objective so that they may pass through the eye's pupil.

Figure 2.8(b) shows the telescope design attributed to Kepler. Kepler's version uses two converging lenses and hence produces an inverted image. This has the advantage that a set of cross-hairs can be placed at the focus, F, so that objects can be accurately placed in the field of view. This is the design typically used on so-called finder telescopes attached to large reflecting telescopes.

Problem 2.5

Use the thin lens equation to show that for both the Galilean design and the Keplerian design the two lenses must be placed a distance $\ell = f_o + f_e$ apart in order for the rays to emerge from the eyepiece

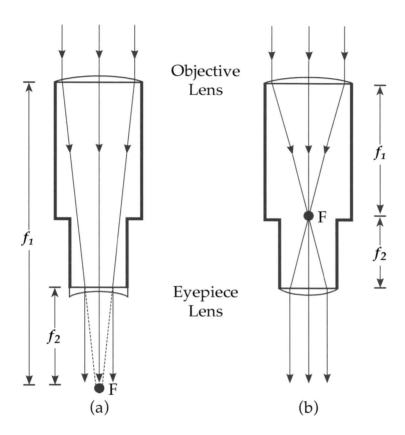

Figure 2.8: Two versions of the refractor telescope. (a) The Galilean design uses one converging and one diverging lens and has a virtual focus. (b) The Keplerian design uses two converging lenses, and has a real focus.

> parallel. Where f_o is the focal length of objective and the f_e is the focal length for the eyepiece.

One way to think of how a Keplerian telescope works is to consider the eyepiece as a magnifying glass used to examine the image formed by the objective. The **angular magnification** m_θ of a telescope is defined to be the angular size of the object divided by the angular size of the image examined with the eyepiece. With a little geometry it is easy to show that

$$m_\theta = -\frac{f_o}{f_e}, \qquad (2.9)$$

where the minus sign is set by convention so that if $m_\theta < 0$ the image is inverted.

Magnification is only one purpose of a telescope. The other perhaps more important function is to gather light. For both designs, the light that passes through the objective finally passes through the eyepiece. The larger the objective the brighter the image appears. The **light-gathering power** of an optical telescope is proportional to the area of the objective lens. Galileo was able to see the moons of Jupiter for the first time because his telescope had sufficient light-gathering power, not because of the magnification of his telescope. If the Galilean moons of Jupiter were brighter they could be resolved as separate from Jupiter with the naked eye.

Galileo did his observations using his eye as a light detector. Today, virtually all astronomical observations are done using a solid state device like a CCD (see Chapter 4) as the detector. Instead of using an eyepiece the detector is placed at the **focal plane** of the objective lens and the image is produced directly on the detector. The focal plane is a plane through the focal point and perpendicular to the optical axis. The simplest telescope is just a lens and a detector.

One problem with refracting telescopes is that since the index of refraction depends on wavelength different colors of light come to a focus at different points. This makes it impossible to obtain a perfect focus. If the lenses are adjusted so that the red light is in focus, the blue light is out of focus and visa versa. This effect is called **chromatic aberration**. Chromatic aberration can be reduced, but not completely eliminated by using **compound lenses**, lenses that are made of two or more pieces of glass with different indices of refraction.

2.1.3 Mirrors

A better way to avoid chromatic aberration is to use curved mirrors to redirect light rays. Most modern astronomical telescopes use mirrors to gather and focus light. Mirrors with spherical surfaces are easy to analyze and fabricate so we will consider them first. Figure 2.9 shows the cross section of a concave spherical mirror of radius r. A ray starting

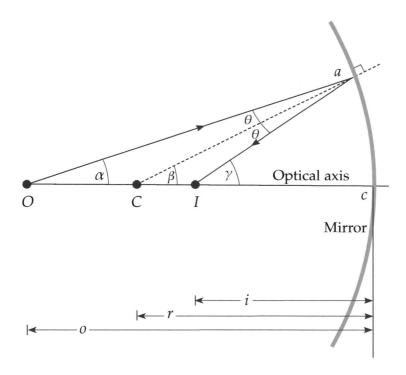

Figure 2.9: Cross section of a concave spherical mirror of radius r. An object is placed at O a distance o from the mirror. A real image is formed at point I a distance i from the mirror.

at point O is reflected from the mirror at point P and passes through the point I. The optical axis of the mirror is defined by the points O, C, and I. Point C is at the center of the spherical surface that forms the mirror. The interior angles of the triangle OCP must add up to π so $\alpha + (\pi - \beta) + \theta_I = \pi$ or $\alpha + \theta_I = \beta$. Likewise for the triangle OIP, $\alpha + (\pi - \gamma) + (\theta_I + \theta_R) = \pi$ or $\alpha + \theta_I + \theta_R = \gamma$. The law of reflection requires that $\theta_I = \theta_R$ so we have the two equations

$$\beta = \alpha + \theta \quad \text{and} \quad \gamma = \alpha + 2\theta,$$

where $\theta \equiv \theta_I = \theta_R$. If we eliminate θ between these two equations we get

$$\alpha + \gamma = 2\beta. \tag{2.10}$$

If the object distance o is large compared to the radius of the mirror, then the incident ray is nearly parallel to the optical axis and

$$\alpha \approx \frac{\overparen{PQ}}{o}, \quad \text{and} \quad \gamma \approx \frac{\overparen{PQ}}{i},$$

where $\overparen{PQ}$ is the arc length from P to Q. The definition of angle also means

$$\beta = \frac{\overparen{PQ}}{r}.$$

Problem 2.6

(a) Show that if the object distance $o \gg r$ then the incident and reflected rays are **paraxial** or approximately parallel to the optical axis.

(b) Use the small angle approximation $\phi \approx \tan \phi$ to show that when the incident and reflected rays are paraxial

$$\alpha \approx \frac{\overparen{PQ}}{o}, \quad \text{and} \quad \gamma \approx \frac{\overparen{PQ}}{i}.$$

Using the relations for α, β, and γ in equation (2.10) gives

$$\frac{1}{i} + \frac{1}{o} = \frac{2}{r}. \tag{2.11}$$

This is identical to the thin lens equation (2.8) with the focal length $f = r/2$. Evidently the focal point of a convex mirror is a distance $r/2$ in front of the mirror. We can use the thin lens equation for mirrors, but with different conventions for the signs of i and f. Concave mirrors behave like converging lenses and have positive focal lengths and form real images if $o > f$, but in this case a positive image distance i means the image is in front of the mirror on the same side of the mirror as the object. Convex mirror act like diverging lenses. They have negative focal lengths, form virtual images, and the negative image diastase means the virtual image is formed behind the mirror.

Problem 2.7

Draw a ray diagram like Figure 2.9, but for a convex mirror. Assume the point O is a distance $o = 2r$ from the mirror. Include the equivalent points O, P, C, I, and Q in your diagram.

2.1.3.1 Simple Reflecting Telescopes

Mirrors work like lenses so you should be able to build a telescope using them. Isaac Newton was the first one to build a reflecting telescope. Figure 2.10 shows his design. His telescope design consists of two mirrors.

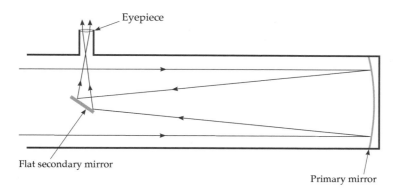

Figure 2.10: Newtonian reflecting telescope. The primary mirror in Newton's original telescope was spherical and suffered from spherical aberration.

Light rays enter the telescope from the left and reflect from a concave spherical **primary mirror** at the right. The primary mirror acts like the objective lens of a refractor and causes the rays to converge. A flat secondary mirror reflects the converging rays out the side of the telescope tube where it passes through an eyepiece.

Newton's design is still used for small amateur telescopes, but it has a significant problem. Rays that strike the mirror far from the optical axis focus to a different point than those that strike close to the optical axis [see Figure 2.11(a)]. This effect is called **spherical aberration.**

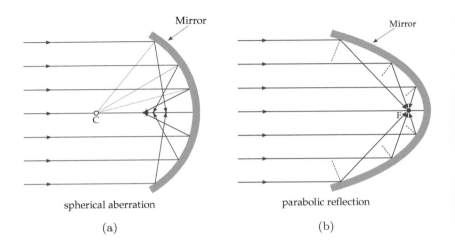

Figure 2.11: (a) Spherical mirrors suffer from spherical aberration. (b) A parabolic mirror eliminates spherical aberration for light rays that are parallel to the optical axis.

Using a parabolic mirror [Figure 2.11(b)] eliminates spherical aberration for rays that originate from points along the optical axis, but introduces other aberrations for rays originating off axis. Designers of modern astronomical telescopes try to eliminate those aberrations by using additional lenses or modifying the shape of the primary and secondary mirrors to compensate for these aberrations. We will explore a few of these designs in Section 2.4. Telescope designs can be quite complex. They may consist of a combination of several mirrors and lenses, but we can always model a telescope as a single objective lens with a single effective focal length.

2.2 IMAGE SCALE

Figure 2.12 shows an Hubble Space Telescope (HST) image of the binary star system Sirius. We know from parallax measurements that the system's distance, $d = 2.64$ pc. If we could measure the angular separation, θ, between Sirius A and Sirius B from the image we could use the small angle formula to determine the physical diameter of orbit, $D = d\theta$. In order to determine θ we have to know how distances in the focal plane

Figure 2.12: A Hubble Space Telescope image of the binary star system Sirius. (Source: NASA/STScI)

of the telescope translate to angles on the sky. The angular distance between two points in the sky divided by the actual distance in the focal plane is called **image scale**[2].

Problem 2.8

Suppose the image scale of the image in Figure 2.12 is 3.04 arcseconds/mm and that the distance between the centers of Sirius A and Sirius B in the focal plane is 2.47 mm. What is the physical distance between Sirius A and B in AU?

It is easy to derive the image scale if we remember some elementary optics. Figure 2.13 shows two stars to the left of a telescope objective lens. The angle on the sky between the two stars is θ. Two rays, one from each star, pass through the center of the objective and form an image of the two stars on the detector. The physical distance between the two stars' images is x. The image scale

$$s = \frac{\theta}{x}. \tag{2.12}$$

The stars are effectively at infinity so the stars' images are produced in the focal plane a distance f to the right of the objective. Rays that

---

[2]The image scale is sometimes called the plate scale. The terminology dates to the days when all astronomical images were taken using photographic plates.

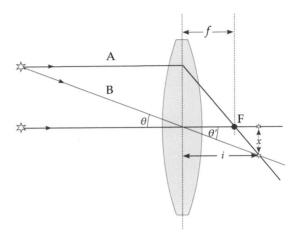

Figure 2.13: A ray tracing diagram showing two stars with an angular separation θ focused by a telescope objective lens. The effective focal length of the telescope is f and the physical distance between the two star images on the focal plane is x.

pass through the center of the lens are undeflected; hence, the angles on either side of the objective are the same. The small angle formula tells us that $\theta = x/f$. Using this in equation (2.12) implies that the image scale is just the reciprocal of the focal length,

$$s = \frac{1}{f}. \tag{2.13}$$

Since we made a small angle approximation, the equation above has units of radians per unit length. If we wish to write the image scale in arcseconds per unit length we have to multiply by the conversion factor, 206,265 arcseconds per radian,

$$s = \frac{206265''}{f}. \tag{2.14}$$

Given the telescope's image scale, s, we can also determine the **field of view** covered by an image taken using a camera. The field of view is just the angular size of the region on the sky covered by the detector. Suppose the detector is height h by width w. The field of view, is simply the image scale multiplied by the physical size of the detector and has units of angle.

Problem 2.9

Suppose a telescope has a focal length $f = 7.2$ m.

(a) What is the image scale in arcseconds per micron?

(b) Suppose the telescope uses a square CCD detector that is 24.6 mm by 24.6 mm. What is the field of view of this telescope-detector system?

2.3 RESOLUTION

As seen from Earth the angular size of all stars except the Sun is exceedingly small. Betelgeuse has one of the largest angular sizes and it is only about 0.06" in diameter. The ability of a telescope to resolve such small angles is limited by diffraction and for ground-based telescopes blurring by the atmosphere.

2.3.1 The Diffraction Limit

Light diffracts as it passes through any opening. This limits any telescope's ability to resolve small angles. The light from a star is diffracted as it passes through the lens of a refracting telescope or reflects from the mirrors of a reflecting telescope. The shape of the mirror or lens is the shape of the opening through which the light diffracts. If the opening is circular then the diffraction pattern of a point source is a bullseye pattern with the highest intensity in the middle and fainter rings around the central maximum. This bullseye pattern is called an **Airy pattern**. The distribution of the intensity of the light is call the **point spread function (PSF)**. The airy pattern PSF is shown in Figure 2.14. The angular distance, θ, between the central maximum and the first diffraction minimum for a circular opening is

$$\theta = \frac{1.22\lambda}{D}, \tag{2.15}$$

where, λ is the wavelength of the light, D is the diameter of the lens or mirror, and θ is in radians.

If two stars have a very small angular separation, their Airy patterns will overlap and we won't be able to resolve them as two separate stars. We take the minimum separation we can resolve to be when the maximum of one star's diffraction pattern is coincident with the other star's

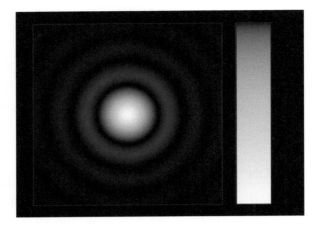

Figure 2.14: A computer simulation of an image of an Airy pattern. The intensities of the outer rings of the pattern have been enhanced to make them more visible. (Source: Wikimedia Commons)

first diffraction minimum (see Figure 2.15). This is called the **Rayleigh criterion**. The minimum angle we can resolve is then given by equation (2.15). This is another reason that astronomers like big telescopes.

Problem 2.10

The Hubble Space Telescope has 2.4-m primary mirror. Its detectors operate in ultraviolet through the infrared part of the spectrum.

(a) What is the minimum angular resolution of the HST at the UV wavelength of 220 nm?

(b) What is the minimum angular resolution at the IR wavelength of 2.4 μm?

2.3.2 Atmospheric Seeing

Because of the atmosphere the resolution of all large ground based optical telescopes is worse than the diffraction limit discussed above. As the light passes through the atmosphere, different parts of the wavefronts pass through air of different densities and hence indices of refraction. The change in the wavefront due to this effect is called **scintillation**. Scintillation distorts the image causing it to change size, shape, position, and brightness. These variations happen over a period of from one to

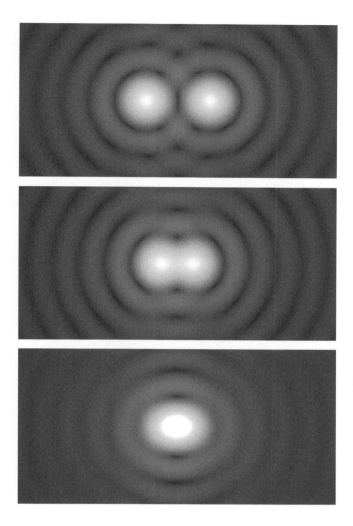

Figure 2.15: The overlapping Airy patterns of two point sources. In the top image the two sources are easily resolved. The middle image shows the two sources when the Rayleigh criterion is met. In the bottom image the two sources are very difficult to resolve. The intensities of the outer rings of the patterns have been enhanced to make them more visible. (Source: Wikimedia Commons)

a thousandth of a second. For any exposure larger than a fraction of a second the airy pattern will be blurred and the star's PSF will be broadened. **Seeing** refers to how much a stellar image is blurred by the atmosphere. Quantitatively, most astronomers define the seeing to be the full width at half maximum (FWHM) of the star's PSF. The best astronomical sites in the world typically have seeing on the order of a few tenths of an arcsecond. At most observatories seeing of an arcsecond is considered to be quite good. Of course, the seeing at any site may vary depending on the atmospheric conditions.

There are ways to reduce the effects of the atmosphere. Of course, one way is to put the telescope in orbit above the atmosphere. Being above the atmosphere, the Hubble Space Telescope is diffraction limited. Its resolution is about a factor of ten better than the best ground based telescopes. Speckle interferometry and adaptive optics are two techniques used at ground based observatories to reduce the effects of atmospheric seeing. These techniques are beyond the scope of this book, but you can learn more about them in Chromey [7, Section 6.6], Bradt [5, Section 5.5], or Tyson [21].

Problem 2.11

Compute the size of the primary mirror of a telescope whose diffraction limited resolution is equal to a seeing of one arcsecond for light in the middle of the visible spectrum ($\lambda \approx 500$ nm).

2.4 TELESCOPE OPTICAL DESIGNS

Telescopes are often characterized by their **focal ratio** sometimes called the **f-number** or **f-ratio**. The f-ratio is defined as the effective focal length of telescope divided by the diameter of the objective lens or primary mirror. Galileo's original telescope had a 37-mm diameter objective lens with a focal length of 980 mm. This gave his telescope an f-ratio of about 26. The f-ratio is usually written as $f/26$.

The first astronomical telescopes were refractors, but virtually all astronomical research telescopes today are reflectors or use both mirrors and lenses. Telescopes that use both refractive and reflective elements are called **catadioptric** telescopes. Refracting telescope designs might use a collection of lens to form an image, but they can all be thought of as a single equivalent objective lens. However, there are a huge number of different designs for reflectors and catadioptric telescopes.

2.4.1 Traditional Designs

Figure 2.16 shows a few of the many traditional designs for reflecting and catadioptric telescopes. Keep in mind that most large research telescopes can be used in a variety of different configurations each having a different f-ratio. For example, the Hale 200-in telescope can be used in a prime focus mode ($f/3.3$), a Cassegrain mode ($f/16$), and a coudé mode ($f/30$). Let's explore the different designs and the advantages and disadvantages of each.

2.4.1.1 Prime Focus Design

Figure 2.16(a) shows the prime focus design. This is the simplest design consisting of nothing more than a parabolic primary mirror. It is just the Newtonian design without the secondary mirror. The detector is simply placed at the focus of the primary mirror. Of course, this only works for large telescopes where the detector doesn't obstruct too much of the aperture. This design's disadvantage is the same as its Newtonian cousin. Light rays that aren't parallel to the optical axis produce optical aberrations in the focal plane. These aberrations can be reduced by using correcting optics mounted in front of the detector. The primary advantage of the prime focus design is that relative to the Cassegrain and Gregorian designs discussed below, the prime focus has the smallest f-ratio. For a given primary mirror size, a small f-ratio means a smaller focal length and hence a larger image scale and field of view.

It is impossible to place a large instrument such as a high resolution spectrometer at the prime focus, but the light can be redirected by mirrors to place the focus at a more convenient location. Figure 2.17 shows a Nasmyth design which is just a modified Cassegrain. In a Nasmyth design, light is reflected off of a diagonal mirror inside the telescope to put the focal plane outside the altitude or declination axis of the telescope. For an alt-az mounted telescope, a platform is attached to the telescope below the focus to support large instruments. For imaging, the Nasmyth design requires a system to compensate for the rotation of the focal plane as the telescope tracks the sky.

Telescopes with an equatorial mount can use a coudé design where several mirrors are used to direct the light from outside the declination axis to place the focal plane at a location that stays fixed as the telescope moves. This configuration is often used for heavy or bulky instruments such as high dispersion spectrographs. The coudé system has a major

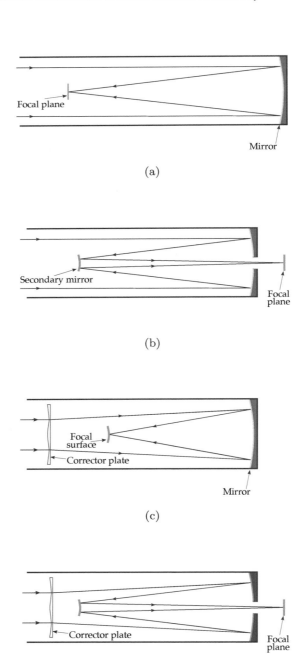

Figure 2.16: Four common telescope optical designs. (a) The Prime focus design. (b) The Cassegrain design. (c) A Schmidt Camera. (d) The Schmidt-Cassegrain design.

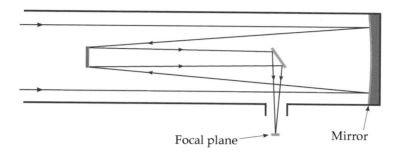

Focal plane ⟶   Mirror

Figure 2.17: Nasmyth focus.

disadvantages for imaging: the field of view is small owing to the large f-number and rotates as the telescope tracks an object across the sky.

2.4.1.2 Cassegrain Design

The Cassegrain design (Figure 2.16(b)) is probably the most common design for research telescopes. The primary mirror reflects light to a secondary mirror which in turn reflects the light back down through a hole in the center of the primary where the light comes to a focus. The location of the focus makes it easy to mount instruments on the back of the telescope. The classical Cassegrain design uses a parabolic primary mirror and a convex hyperbolic secondary mirror which lengthens the focal length of the system to be several times that of the primary mirror alone. Unfortunately, this design is subject to large aberrations so that light rays that converge off the optical axis don't come to a sharp focus. These aberrations are reduced significantly by using a hyperbolic primary mirror and adjusting the shapes of the primary and secondary to reduce optical aberrations. This type of Cassegrain is called a Ritchey-Chrétien design and is quite common. Many large research telescopes can be configured to either mount a detector at the prime focus or to place a secondary mirror near the focus to turn the telescope into a Cassegrain design.

2.4.1.3 Catadioptric Designs

Figures 2.16(c) and (d) show two designs optimized to produce a large field of view. They both use both reflective and refractive optics. In the **Schmidt Camera** design, light passes through a refractive element called a corrector plate, reflects off of a spherical primary mirror, and comes to focus on a *curved* focal surface inside the telescope. The corrector plate removes spherical aberration. In the first Schmidt cameras, a photographic plate was carefully forced into a curved shape and placed at the focal surface. Schmidt cameras can produce fields of view as large as eight degrees.

The disadvantages of the Schmidt design is the location and curvature of the focal surface. The **Schmidt-Cassegrain design** shown if Figure 2.16(d) eliminates these problems by using a curved secondary mirror to reflect light back through the primary to the Cassegrain focus. This has the added benefit of allowing the designer to shape the secondary to not only eliminate the curvature of the focal surface, but to further reduce other optical aberrations.

2.4.2 Modern Telescope Designs

Many modern telescopes are one of the traditional designs described above. The Hubble Space Telescope for example is a Ritchey-Chrétien design. The Hale 200-inch telescope completed in 1948, the world's largest telescope until 1976, was built to operate in a prime focus mode, a Cassegrain mode, and a coudé mode. Beginning in the early 1970s designers began to explore more unconventional designs. The Multiple Mirror Telescope (MMT) completed in 1979 was one of the first. The MMT employed six mirrors, each with a diameter of 1.8 m. This gave it the equivalent light gathering area of a 4.5-m telescope. At the time of its construction essentially all major telescopes used an equatorial mount. The MMT and all large optical telescopes since the MMT have been built with alt-az mounts.

The larger the primary mirror of a telescope, the greater the telescope's light gathering power. The Hale telescope's 200-in primary mirror is made of Pyrex glass and weighs about 14 tons. It had to be made this massive to maintain the shape of its reflective surface. Mirrors any larger than this will deform under their own weight as they are moved to point the telescope. The twin Keck 10-m telescopes built in the 1990s avoided this problem by constructing the primary mirror out of 36 smaller hexagonal mirrors. The individual mirrors are actively controlled to maintain

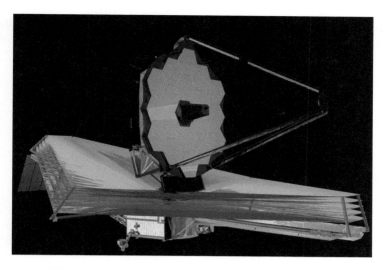

Figure 2.18: The James Webb Space Telescope. Each hexagonal segment in the primary mirror is 1.32 m in diameter. The total length of the sun shield is about 20 m. (Source: NASA)

the correct shape to within a few nanometers as the telescope is pointed. This sort of segmented design coupled with actively controlling the primary's shape is now common on the world's largest telescopes.

The James Webb Space Telescope (JWST) shown in Figure 2.18 is a particularly innovative design. The JWST is the successor to the HST and is currently scheduled to launch in 2021. The primary mirror is made up of 18 hexagonal segments and has an equivalent diameter of 6.5 m. It will be put into at the Earth-Sun L2 point[3]. It is optimized to work in the long-wavelength visible to the mid-infrared. The 16 segments will unfold after launch and be actively controlled and adjusted periodically to maintain optimum image quality.

SUPPLEMENTARY PROBLEMS

Problem 2.12 Derive the law of reflection from Fermat's Principle.

Problem 2.13 Derive the Snell's Law from Fermat's Principle.

---

[3]L2 is one of the five Lagrange points. It is a point fixed with respect to a line from the Sun to the Earth. It is 1.5×10^6 km away from the the nighttime side of the Earth.

Problem 2.14 Calculate the angular size of orbits of Galilean satellites as seen from Earth and compare these angles to the angular resolution of the human eye.

CHAPTER 3

Measuring Light

Virtually everything we know about the cosmos comes from the study of electromagnetic radiation. We have sent spacecraft to a few nearby objects in our solar system to analyze or bring back samples. We routinely detect cosmic rays—high energy particles originating in spectacular astrophysical events like supernovae. But most of what we know comes from measuring the intensity of electromagnetic radiation.

3.1 FROM THE STARS TO OUR DETECTOR

We commonly refer to the brightness of a star, but what exactly do we mean by brightness? Astronomers quantify the brightness of a celestial object by specifying the **flux** or **flux density** F received from the star.[1] The flux is the total energy from a source that crosses a unit area per unit time:

$$F = \frac{\text{energy}}{\text{time} \cdot \text{area}}.$$

The total power output from an object is called the **luminosity**. Suppose a star of luminosity L is emitting its light isotropically (the same in all directions). The star's flux measured at distance r from the star is just the luminosity of the star divided by the total area $4\pi r^2$ of a sphere centered on the star:

$$F = \frac{L}{4\pi r^2}. \tag{3.1}$$

Equation (3.1) is the famous **inverse square law of radiation**.

---

[1] What astronomers commonly call flux or flux density physicists typically call **irradiance**. Physicists also use the term **radiant flux** instead of **luminosity** to mean the total power output of a source.

DOI: 10.1201/9781003203919-3

Problem 3.1

The luminosity of the star Vega is 1.5×10^{28} W and it is at a distance of 7.68 pc. What is flux of Vega at Earth in W/m^2?

The detectors we use at telescopes to measure the brightness of a star measures the power delivered to a sensor by electromagnetic radiation or the number of photons collected by the detector. Even the retina of our eye works as a photon counter. Astronomical objects are usually quite faint, so the primary function of a telescope is to gather light. The simplest telescope is just a converging lens. To measure the brightness of a star, we would simply point the lens at the star and place a sensor in the focal plane to capture all the radiation in the star's image. If the flux from an object is F and the area of the collecting area of the telescope is A then the power that falls on the detector is

$$P = AF.$$

Note that the larger the area of the objective lens or primary mirror, the greater the power focused on the detector. This makes it clear why astronomers continue to crave telescopes with ever larger apertures. Currently the largest functional telescopes in the world have apertures diameters on the order of 10 m.

Problem 3.2

Suppose you are observing Vega using the Keck 10-m telescope. What is the power at the detector assuming the primary mirror is circular and perfectly reflecting? Ignore things like atmospheric extinction.

When F and L above are the total flux and power output integrated over *all* wavelengths and measured by a perfect detector they are referred to as the **bolometric flux** F_{bol} and **bolometric luminosity** L_{bol}. Measuring the bolometric flux is extremely difficult. There are no detectors capable of measuring power over the entire electromagnetic spectrum. In practice we measure different parts of the spectrum with different detectors. The **monochromatic flux** is the flux per unit wavelength F_λ or frequency F_ν.[2] Integrating the monochromatic flux over the

---

[2]Physicists usually call the monochromatic flux the spectral irradiance.

entire spectrum gives the bolometric flux density,

$$F_{\text{bol}} = \int_0^\infty F_\lambda d\lambda = \int_0^\infty F_\nu d\nu. \tag{3.2}$$

The monochromatic flux is usually what we mean when we refer to the spectrum of the light. We call both F_λ and F_ν the monochromatic flux. They are related, but they have different units and are not numerically equal.

Problem 3.3
Use that facts that $F_\lambda d\lambda$ is the flux in the range of wavelengths between λ and $\lambda + d\lambda$ and that $F_\nu d\nu$ is the flux in the range of frequencies between ν and $\nu + d\nu$ to show that

$$\lambda F_\lambda = \nu F_\nu. \tag{3.3}$$

Hint: Don't forget that $\lambda\nu = c$ for light.

The units F_λ are energy per unit time per unit area per unit wavelength. Its units could be written as W m^{-3}, but this is almost never done. Since we will often be working with wavelengths in nm it is more convenient to use the units W m^{-2} nm^{-1}. Typical units for F_ν are W m^{-2} Hz^{-1}. Another unit called the janskys (Jy) is used extensively by radio astronomers. Kark Jansky essentially created the field of radio astronomy when he discovered radio waves emanating from the center of our galaxy; 1 Jy = 10^{-26} W m^{-2} Hz^{-1}. The monochromatic flux of an astronomical object defines its spectrum. Figure 3.1 shows the spectrum of stars of two different spectral classes.

In practice, we can't measure the bolometric or monochromatic fluxes directly. Different kinds of detectors are sensitive to light of different wavelengths and any one detector's sensitivity varies as a function of wavelength. The problem is even worse for ground based telescopes. The atmosphere attenuates some wavelengths more than others. We'll see how to correct for atmospheric effects in Chapter 6. If we ignore effects of the atmosphere for the time being, then the flux measured by a telescope is

$$F = \int_0^\infty R(\lambda)F_\lambda d\lambda, \tag{3.4}$$

where $R(\lambda)$ is called the **response function** of the detector-telescope

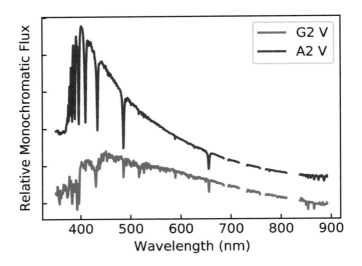

Figure 3.1: The monochromatic flux F_λ or spectra of a G2 and an A2 main sequence stars. (Data from Silva and Cornell [18].)

system. The response function specifies the fraction of the the flux of wavelength λ that is registered by the detector-telescope system. The power P measured by the detector-telescope system is

$$P = A \int_0^\infty R(\lambda) F_\lambda d\lambda. \tag{3.5}$$

Astronomers have devised sets of agreed-upon response functions that define several standard photometric systems. Having well-defined response functions allows astronomers to compare observations made with different telescope-detector systems. They use filters to make sure the response function of their system matches one of the standard systems. The response of a system depends on the reflectivity or transmissivity of the telescope optics T_o, the detectors efficiency at detecting light of wavelength $e(\lambda)$, and the transmissivity of the filter $\phi(\lambda)$

$$R(\lambda) = T_o e(\lambda) \phi(\lambda). \tag{3.6}$$

The transmissivity of the filter $\phi(\lambda)$ is called the **filter function**. The **quantum efficiency** $e(\lambda)$ is the fraction of incoming photons that are actually detected. The transmissivity of the optics T_o usually doesn't depend on wavelength over the bandpass of the filter.

Problem 3.4

The star Vega is one of the brightest in the night sky. In this problem you will estimate the power from Vega measured by a detector on a small telescope.

(a) Use equations 3.5 and 3.6 to show that

$$P_D = AT_o \int_0^\infty e(\lambda)\phi(\lambda)F_\lambda d\lambda.$$

(b) The star Vega's monochromatic flux above the atmosphere at 550 nm is approximately $3.6 \times 10^{-11} \text{W m}^{-2} \text{nm}^{-1}$. Estimate the power from Vega on a detector attached to a 0.61-m telescope. Assume the optics transmissivity is 0.7 and an average quantum efficiency is 0.5 at 550 nm. Assume the filter function is a step function with a maximum of 0.5 centered on 550 nm and 89 nm wide. In this case, include the effect of the atmosphere by assuming the atmospheric transmissivity is about 0.8.

3.2 THE MAGNITUDE SYSTEM

Long before the physics of electromagnetic radiation was understood, astronomers were classifying stars by their apparent brightness. In the second century BC a Greek astronomer named Hipparchus of Nicea compiled a catalog of the position and brightness of nearly a thousand stars. He knew nothing of luminosity or flux. Instead he classified stars by their brightness as seen by the unaided eye. He assigned a number or magnitude to each star. Magnitude one stars were the brightest in the night sky and magnitude six stars were the faintest he could see. The eye has a logarithmic response to light energy so that five magnitudes corresponds to a factor of about 100 in flux. In 1856, Norman R. Pogson of Oxford tied the magnitude system to flux by proposing that five magnitudes corresponds to a factor of *exactly* 100 in flux. The magnitude m is therefore defined as

$$m = -2.5 \log(F) + C. \tag{3.7}$$

The constant C defines the **zero point** of the magnitude system through the equation $C = -2.5 \log(F_0)$, where F_0 is the flux density from a $m = 0$ star. Astronomers originally set the zero point to closely approximate Hipparchus' magnitudes by assigning the star Vega to be $m = 0$. This essentially makes Vega the standard against which all the other stars are measured. The problem with this system is that Vega is slightly variable, by about 0.03 magnitudes, and has an atypical infrared spectrum. Today the flux for zero magnitude is set by averaging the magnitudes of many stars or by calibrating the scale in absolute flux units. In practice we always measure the magnitude of a star by comparing its flux to a standard star with known magnitude. The difference in the magnitude of the two stars $m_1 - m_2$ is independent of the constant C,

$$m_1 - m_2 = -2.5 \log \left(\frac{F_1}{F_2} \right), \tag{3.8}$$

where F_1 and F_2 are the measured fluxes from stars with magnitudes m_1 and m_2 respectively.

Problem 3.5

Use equation (3.7) to prove equation (3.8).

Problem 3.6

Solve equation (3.8) for the ratio F_1/F_2 to show that

$$\frac{F_1}{F_2} = 10^{-0.4\Delta m}, \tag{3.9}$$

where $\Delta m = m_1 - m_2$.

3.2.1 Absolute Magnitude and Distance Modulus

Astronomers seem to love the magnitude system. They use magnitudes to specify luminosities and distances as well as flux densities. The **absolute magnitude** M is defined as the the apparent magnitude a star would have if observed from a distance of 10 parsecs (pc). Using the inverse square law [equation (3.1)] and the definition of magnitude [equation (3.7)] you can show that

$$M = -2.5 \log(L) + C', \tag{3.10}$$

where C' is a constant related to C. Specifying the absolute magnitude of an object is equivalent to specifying its luminosity.

Problem 3.7
The Sun has an absolute bolometric magnitude of 4.74, what is its apparent magnitude seen here on Earth?

The **distance modulus** μ is defined as the difference between a celestial objects apparent and absolute magnitudes and it depends only on the object's distance

$$\mu \equiv m - M = 5\log(r) - 5, \tag{3.11}$$

where r is the distance to the object *in parsecs*.

Problem 3.8
Use the definition of absolute magnitude and the inverse square law of radiation, equation (3.1), to prove equation (3.11).

Solving equation (3.11) for the distance gives

$$r = 10^{(\mu+5)/5}. \tag{3.12}$$

Here again remember that r is given in parsecs.

Problem 3.9
A type Ia supernova (SNIa) is observed to have an apparent magnitude of 12.1 at its brightest. Most SNIa have an absolute magnitude of $M = -19.3$ at maximum light.
(a) What is the distance modulus for this supernova?
(b) What is the supernova's distance in Mpc?

3.2.2 Standard Filter Systems

In practice, we never measure the bolometric magnitude directly, because our detectors are only sensitive to light in a finite wavelength band. We can also get a good deal of information from knowing the color of the object. In order to compare observations made at different observatories with different detectors, astronomers have to agree on standard response functions. One of the first and still one of the most commonly used is the Johnson–Cousins UBVRI system. Figure 3.2 shows the response

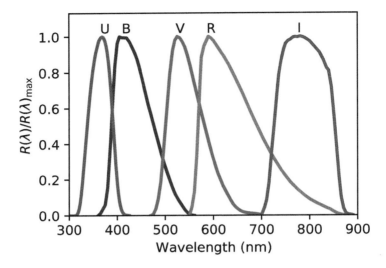

Figure 3.2: Response functions for the Johnson–Cousins UBVRI photometric system.

functions for the UBVRI system.[3]

Table 3.1 lists the bandpasses for the Johnson–Cousins UBVRI system with the approximate central wavelength and width of each response function. The peak response of the U bandpass is in the ultraviolet part

Table 3.1: Johnson–Cousins UBVRI system response function effective wavelengths and widths. (Data from Bessell[1].)

Bandpass	U	B	V	R	I
λ_{eff} (nm)	366	436	545	641	798
$\Delta\lambda$ (nm)	65	89	84	158	154

of the spectrum. The B response peaks in the blue and has response similar to photographic emulsions. The V filter peaks in the green. The V stands for visual since the V-band's response is close to that of the hu-

---

[3]Johnson originally defined the R and I bands differently. The curves shown here are for Cousins' modifications to the original system. Sometimes you will see the Johnson bands designated R_J and I_J and the Cousins bands as R_C and I_C.

man eye. The R and I bands peak in the red and the near infrared part of the spectrum. Astronomers work hard to make sure their telescope-detector systems match the response functions.

The magnitudes for this system are designated by capital letter. For example, the magnitude for the U band is

$$U = -2.5 \log(F_U) + C_U, \tag{3.13}$$

where C_U sets the U-band zero point. F_U is

$$F_U = \int_0^\infty R_U(\lambda) F_\lambda d\lambda, \tag{3.14}$$

where $R_U(\lambda)$ is the U-band response function. The zero point constants for U, B, V, R, and I are all different and were originally chosen so that Vega had zero magnitude in all bands. As discussed above, the zero-point standards are chosen differently now, but Vega's magnitude is still close to zero using these new standards. The extended UBV system is described in more detail in Chapter 6.

3.2.3 Color Indices

If we were able to measure a stars monochromatic flux at *all* wavelengths we could determine its bolometric magnitude,

$$m_{\text{bol}} = -2.5 \log(F_{\text{bol}}) + C_{\text{bol}}, \tag{3.15}$$

but this would require a complete spectrum of the star and that is hard to measure. One way to approximate the whole spectrum is to measure the relative fluxes through our set of standard filters. This is a sort of crude spectroscopy, but it does gives us the color of the star. More blue light than red makes the star look bluish. A quantitative measure of the color is called the **color index**. It is just the difference between the magnitudes in two different band passes, and gives the ratio of the fluxes in these two bandpasses. For example the $B - V$ color index is

$$B - V = -2.5 \log \left(\frac{F_B}{F_V} \right) + C_{B-V}, \tag{3.16}$$

where $C_{B-V} \equiv C_B - C_V$. A large $B - V$ color index implies that $B > V$ so the star is less blue (larger B magnitude) than one with a small $B-V$. Hence, a star with large $B - V$ color index is redder than one with a small $B - V$.

Color indices are so useful in fact that most standard star catalogs list the V magnitude and the color indices for stars rather than the individual magnitudes. The color indices of a star are also independent of its distance so that

$$B - V = M_B - M_V, \tag{3.17}$$

where M_B and M_V are the absolute magnitudes of the star in the B band and V band respectively.

Problem 3.10
Prove equations (3.16) and (3.17).

If the color indices of a star are known then in principle we should be able to approximate its bolometric magnitude. Astronomers have used a combination of stellar-spectra modeling and observations to derive a **bolometric correction** BC that is used to determine the bolometric magnitude of a star from its V magnitude,

$$BC = m_{\text{bol}} - V. \tag{3.18}$$

The bolometric correction depends on the spectral and luminosity class of a star. Typically astronomers use the color indices or spectrum of the star to determine its type and then apply the bolometric correction to the V magnitude to determine the star's bolometric magnitude.

In Chapter 6, we will discuss another photometry system originally developed for the Sloan Digital Sky Survey. This system consists of five filter bands u', g', r', i', and z' that cover the near UV to the near IR part of the spectrum. One can define color indices for this system, usually referred to as the ugriz system, in the same way color indices are defined for the UBVRI system. For a detailed discussion of this an other standard photometric systems see Bessell [1].

3.3 MEASURING MAGNITUDES

Detectors used for measuring light can be classified into two broad categories: photon counting detectors and detectors for measuring power. Magnitudes are defined in terms of flux densities. How do we translate from measurements made at the telescope to magnitudes in a standard system? We start by noting that we typically measure a star's magnitude

m by comparing it to a standard star whose magnitude m_s is known. Using equation (3.8) gives

$$m = m_s - 2.5 \log \left(\frac{F}{F_s} \right), \tag{3.19}$$

where F and F_s are the unknown star's flux and standard star's flux respectively. These fluxes depend on the response functions for the standard bandpass being used [see equation (3.4)]. If the detector measures power, then since $P = FA$, we can write equation (3.19) in terms of the measured powers of the unknown and the standard stars,

$$m = m_s - 2.5 \log \left(\frac{P/A}{P_s/A} \right) = m_s - 2.5 \log \left(\frac{P}{P_s} \right). \tag{3.20}$$

When using this method, measuring the power is equivalent to measuring the flux.

3.3.1 Counting Photons

Unfortunately, this isn't quite what we want. Most modern detectors that work in the in ultraviolet through near infrared parts of the spectrum are photon counters—they don't measure power. The photon count rate

$$\dot{n} = A \int_0^\infty R_p(\lambda) \dot{n}_\lambda d\lambda, \tag{3.21}$$

where $R_p(\lambda)$ is the **photon response function** and $\dot{n}_\lambda$ is the **monochromatic photon flux**. Here A is again the collecting area of the telescope. The monochromatic photon flux is the number of photons of wavelength λ that are counted each second per unit of telescope collecting area. We can relate the monochromatic photon flux to the monochromatic flux by dividing the energy flux by the energy of a single photon $E_{\text{photon}} = hc/\lambda$,

$$\dot{n}_\lambda = \frac{F_\lambda}{hc/\lambda}. \tag{3.22}$$

This allows us to rewrite the count rate in terms of the flux density,

$$\dot{n} = A \int_0^\infty R_p(\lambda) \frac{\lambda}{hc} F_\lambda d\lambda. \tag{3.23}$$

Comparing equation (3.4) for the measured flux and (3.23) we see that if $R(\lambda) \propto \lambda R_p(\lambda)$, then $\dot{n} \propto F$ and

$$m = -2.5 \log (\dot{n}) + C. \tag{3.24}$$

All we need to do is to make sure that our photon counting detector gives magnitudes in the standard system is to adjust the photon response function with filters to make sure $R_p(\lambda) \propto R(\lambda)/\lambda$.

Problem 3.11
Show that if $R_p(\lambda) \propto R(\lambda)/\lambda$ then $m = m_s - 2.5 \log (\dot{n}/\dot{n}_s)$.

3.3.2 Uncertainties in Brightness Measurements

Even with a perfect, noise-free detector we cannot measure the brightness of a star with absolute certainty. The reason for this is that the emission and detection of photons is probabilistic. To clarify this let's imagine a noiseless detector is used to count the number of photons from a star in a period of one second. There is no uncertainty in the number of photons registered by the detector. Nevertheless, if we repeat the measurement we will get a different number. A different number of counts arises because each measurement is a single sample drawn from a Poisson distribution (see Appendix B). This means that if we count n photons with our detector, the best estimate of the uncertainty in the number of counts $\delta n = \sqrt{n}$. If we measure a star and get $n = 10\,000$ then $\delta n = 100$ and the fractional uncertainty in our measurement is $\delta n/n = 0.01$ or 1%. We define the **signal-to-noise ratio** to be the inverse of the fractional uncertainty,

$$\text{SNR} = \frac{n}{\delta n}. \tag{3.25}$$

In the example above where $n = 10\,000$ and $\delta n = 100$, SNR $= 100$. Of course real photon detectors aren't noise-free. We have to add the detector noise to the photon counting noise. If the detector noise σ_{det} isn't correlated with the counting noise, we add the two sources of noise in quadrature. The counting noise is $\sqrt{n}$ so

$$\delta n = \sqrt{n + \sigma_{\text{det}}^2}, \tag{3.26}$$

where σ_{det} is expressed in equivalent photon counts.

Problem 3.12
Suppose we are using a photon counting detector with $\sigma_{\text{det}} = 90$ photons to measure the brightness of a star. Using this detector we count 55 232 photons.

> (a) What is the uncertainty in the measurement?
> (b) What is the signal-to-noise ratio for the measurement?

Equation (3.24) tells us that the brightness of a star is related to the photon count rate $\dot{n}$, not the raw number of counted photons n. But, if we expose our detector to the starlight for a time t then $n = \dot{n}\,t$. We can rewrite equation (3.24) in terms of photon counts,

$$m = m_s - 2.5\log\left(\frac{n/t}{n_s/t_s}\right), \tag{3.27}$$

where t is the exposure time for the measured star and t_s is the exposure time for the standard star. By using equation (B.27) for error propagation from Appendix B we can derive an equation for the uncertainty in magnitude,

$$\delta m = \frac{2.5}{\ln(10)}\frac{\delta n}{n} \approx \frac{\delta n}{n}, \tag{3.28}$$

where we have assumed that the uncertainties in the standard star magnitude, the count rate of the standard star, and the exposure time are negligible. The uncertainty in the magnitude is approximately equal to the *fractional* uncertainty in the number of photons or is inversely proportional to the signal-to-noise ratio,

$$\delta m \approx \frac{1}{\text{SNR}}. \tag{3.29}$$

This means that if we make a measurement with an uncertainty of 0.1 magnitudes, the fractional uncertainty in counts is 10% and the signal-to-noise ratio is 10.

Problem 3.13
Use the equations for error propagation in Appendix B to prove equation (3.28).
Hint: Remember $\log(x) = \frac{\ln(x)}{\ln(10)}$.

3.4 ESTIMATING EXPOSURE TIMES

When preparing to measure the brightness of an object we often know the approximate magnitude, and we need to estimate the exposure time to use to get a certain accuracy in our final measurement. In order to calculate an exposure time we need to have at least a rough estimate of the count rate produced on the detector by a star of a given magnitude. The count rate depends on the collecting area of our telescope, the response function of the telescope-detector system, and of course the magnitude of the star. If we knew the count rate for a zero magnitude star is $\dot{n}_0$, then the count rate for a star of magnitude m is

$$\dot{n} = \dot{n}_0 10^{-m/2.5}. \tag{3.30}$$

Hence, if we know the count rate for a zero magnitude star we can always calculate the rate for any other magnitude.

Problem 3.14
Use equation (3.24) to derive equation (3.30).

The count rate $\dot{n}_0$ depends on the flux from a zero magnitude star. Table 3.2 gives the effective monochromatic flux for zero-magnitude A0 stars in the Johnson-Cousins UBVRI photometric system. The **effective monochromatic flux** is defined as

$$F_{\lambda\text{eff}} = \frac{\int_0^\infty R(\lambda)F_\lambda d\lambda}{\int_0^\infty R(\lambda)d\lambda} \tag{3.31}$$

and is just the average of the monochromatic flux over the bandpass. The effective monochromatic flux depends on the spectrum of the star

Table 3.2: Effective monochromatic fluxes and photon count rates for a zero-magnitude A0 in the Johnson-Cousins UBVRI photometric system. The effective wavelengths and widths are the same as those Table 3.1. The flux data are derived from Bessell *et al.* [2].

Bandpass	U	B	V	R	I
λ_{eff} (nm)	366	436	545	641	798
$\Delta\lambda$ (nm)	65	89	84	158	154
$F_{\lambda\text{eff}}$ ($\times 10^{-13}$ W m^{-2} nm^{-1})	418	632	363	218	113
$\dot{n}_{\lambda\text{eff}}$ ($\times 10^6$ s^{-1} m^{-2} nm^{-1})	77	139	100	70	45

so objects whose spectra differs from an A0 star will have different $F_{\lambda\text{eff}}$, but the data in the table will still allow us to calculate an *approximate* photon count rate.

Equation (3.23) allows us to compute the photon count rate from the monochromatic flux. By using the definition of $F_{\lambda\text{eff}}$ in equation (3.23) and the fact that $R_p(\lambda) \propto R(\lambda)/\lambda$ we can show that

$$\dot{n} = A \frac{F_{\lambda\text{eff}}}{hc/\lambda_{\text{eff}}} \int_0^\infty R_P(\lambda)d\lambda.$$

If we define the effective monochromatic photon flux

$$\dot{n}_{\lambda\text{eff}} = \frac{F_{\lambda\text{eff}}}{hc/\lambda_{\text{eff}}}, \tag{3.32}$$

then

$$\dot{n} \approx A\,\dot{n}_{\lambda\text{eff}} \int_0^\infty R_P(\lambda)d\lambda. \tag{3.33}$$

Values for $\dot{n}_{\lambda\text{eff}}$ for a zero magnitude star are also listed in Table 3.2.

We only want a rough estimate so we can make some approximations to evaluate the integral. Equation 3.6 shows that the response function depends on the the transmissivity of the optics T_o, the efficiency of the detector $e(\lambda)$, and the filter function $\phi(\lambda)$. If we assume the detector efficiency is roughly constant across the bandpass, and that the filter function is approximately a step function centered on λ_{eff} with width $\Delta\lambda$ and with a height ϕ', then

$$\int_0^\infty R_P(\lambda)d\lambda \approx T_o e' \phi' \Delta\lambda,$$

where e' is the detector efficiency in the bandpass. Of course, if we are working with a ground-based telescope we should also include a transitivity factor for the atmosphere. We will assume here we are using a space-based telescope. Here R_p' is a constant equal to the fraction of the incoming photons that make it through the optics and are detected by the detector. We can now write equation (3.33) for the count rate in a more useful form

$$\dot{n} \approx A T_o e' \phi' \dot{n}_{\lambda\text{eff}} \Delta\lambda. \tag{3.34}$$

This equation allows us to estimate the photon count rate if we know its magnitude, the characteristics of our telescope-detector system, and which standard photometric filter we are using.

Let's work through an example to clarify how we use equation (3.34). Suppose we wish to measure a $V = 9.5$ star with a 0.41-m telescope. We will assume that the transmissivity of the optics is $T_o \approx 0.8$, that the detector efficiency is $e' \approx 0.5$, and that the transmissivity of the filter is $\phi' = 0.6$. Using the values of $\dot{n}_{\lambda\mathrm{eff}}$ and $\Delta\lambda$ for the V-filter in Table 3.2 along with the values of T_o, e', and ϕ' in equation (3.34) we get the photon count rate $\dot{n}_0$ for a $V = 0$ star,

$$\dot{n}_0 \approx \left[\pi\left(\frac{0.41}{2}\,\mathrm{m}\right)^2\right](0.8)(0.5)(0.6)\left(100\times10^6\frac{1}{\mathrm{s\,m^2\,nm}}\right)(84\,\mathrm{nm})$$
$$\approx 2.7\times10^8\,\mathrm{s^{-1}}.$$

The first factor in brackets is the computed collecting area of the 0.41-m telescope. We now use equation (3.30) to compute the count rate for a $V = 9.5$ magnitude star and get

$$\dot{n} = (2.7\times10^8\,\mathrm{s^{-1}})10^{-9.5/2.5} = 4.2\times10^4\,\mathrm{s^{-1}}.$$

Problem 3.15

Compute the expected photon count rate from a $B = 6.5$ star using a 1-m telescope. The transmissivity of the optics is $T_o = 0.8$, and the filter transmissivity is $\phi' = 0.5$. Most astronomical detectors have poor blue efficiency so assume $e' = 0.3$.

We now have a way to estimate the detector's photon count rate. The next step is to decide how accurately we would like to know the result. Longer exposure times allow us to collect more photons and hence reduce the uncertainty in the measurement. Suppose we want to measure the magnitude of our $V = 9.5$ star with an accuracy of 0.01 magnitudes. Equation (3.29) tells us this measurement will require a SNR $= 100$ or

$$\frac{n}{\delta n} = 100.$$

If we assume a noiseless detector, $\sigma_{\mathrm{det}} = 0$, then $\delta n = \sqrt{n}$ and

$$\frac{n}{\delta n} = \sqrt{n} = 100,$$

so $n = 10\,000$. Given that $n = \dot{n}\,t$ where t is the exposure time,

$$t = \frac{n}{\dot{n}} = \frac{1.0\times10^4}{4.2\times10^4\,\mathrm{s^{-1}}} = 0.24\,\mathrm{s}.$$

We need at least a 0.24 s exposure time to measure our star with a SNR $= 100$ or an accuracy of 0.01 magnitudes.

Problem 3.16
Use the count rate of Problem 3.15 to estimate the exposure time for the $B = 6.5$ if we require an accuracy of 0.005 magnitudes.

If the detector isn't noiseless, and no detector is, the exposure time calculation is more complicated. In terms of the signal to noise ratio

$$\text{SNR} = \frac{\dot{n}\,t}{\sqrt{\dot{n}\,t + \sigma_{\text{det}}^2}}. \tag{3.35}$$

Solving this equation for t gives us an equation for the exposure time in terms of the photon count rate and the desired signal-to-noise ratio,

$$t = \frac{\text{SNR}^2}{\dot{n}} \left(\frac{1}{2} + \sqrt{\frac{1}{4} + \left(\frac{\sigma_{\text{det}}}{\text{SNR}} \right)^2} \right). \tag{3.36}$$

Problem 3.17
(a) Use equation (3.35) to derive equation (3.36). Hint: Remember, exposure times are always positive.
(b) Show that if $\sigma_{\text{det}} \ll \text{SNR}$ that equation (3.36) reduces to the equation for a noiseless detector

$$t = \frac{\text{SNR}^2}{\dot{n}}. \tag{3.37}$$

Problem 3.18
Compute the exposure time for the $B = 6.5$ star of Problem 3.15 assuming a detector noise $\sigma_{\text{det}} = 90$.

SUPPLEMENTARY PROBLEMS

Problem 3.19 We ignored the effects of the atmosphere in this chapter, but suppose the atmosphere absorbed and scattered 20% of the light from a star so that the atmospheric transmissivity $T_A = 80$.

(a) Redo the calculation of Problem 3.18, but this time taking into account absorption and scattering by the atmosphere.

(b) What is the atmospheric extinction in magnitudes, in other words by how much is the magnitude of a star increased because of absorption and scattering by the atmosphere?

Problem 3.20 Suppose we wanted to measure the V magnitude of a star with an accuracy of 0.01 magnitudes. We know that the approximate magnitude of the star is $V = 16.5$. We are using a telescope with a 2.4-m primary mirror. Assume the transmissivity of the telescope optics is $T_o \approx 0.8$, the detector efficiency in the V-band is $e' \approx 0.9$, and the transmissivity of the V filter is $\phi' \approx 0.8$. You may ignore atmospheric extinction. What approximate exposure time would we need to make the measurement?

Problem 3.21 Suppose we also wanted to measure the $B - V$ color index of the star of Problem 3.20 with an accuracy of 0.03 magnitudes. What exposure time will we need for the B-magnitude measurement? Suppose we have already made the V measurement with an accuracy 0.01 magnitudes, that T_o is the same in the B-band as the V-band, but that $e' \approx 0.5$ and $\phi' \approx 06$. Assume you know $B \approx 16.0$. You may again ignore atmospheric extinction.

CHAPTER 4

Charge-Coupled Devices

Hipparchus and Galileo used their eyes to make astronomical observations, but human eyes aren't very good light detectors for astronomy. They don't have a linear response, their sensitivity is limited, and there isn't a good way to quantify their output. Astronomers began using photographic plates as light detectors in the middle of the 19th century. By the early 20th century photographic plates were used for photometry and spectroscopy as well as imaging. Photographic plates allowed for long exposure times so they could detect low light levels, but like the human eye, they don't have a linear response. By 1950 astronomers began using photomultipliers for photometry, but most spectroscopy and imaging still used photographic plates. In the 1970s, astronomers began to experiment with solid-state arrays for both imaging and spectroscopy. The most common type of solid-state array used in astronomy today is a charge-coupled device (CCD). CCDs were invented by Willard Boyle and George Smith in 1969 at Bell Labs who originally intended them to be used as memory storage devices. Bell Labs engineers quickly realized that they also made good low-intensity light detectors. CCDs are far more efficient at detecting light than either photographic plates or photomultipliers, have a linear response, and have a broad spectral response. They are by far the most common light detectors used by astronomers at visual and near-infrared wavelengths.

DOI: 10.1201/9781003203919-4

4.1 LIGHT DETECTION

CCDs are two-dimensional arrays of photosensitive cells called **picture elements** or **pixels**. Most CCDs used in astronomy have several million pixels. The pixels are anywhere from 5 to 20 μm in size. Each pixel is essentially a metal-oxide-semiconductor (MOS) capacitor. Figure 4.1

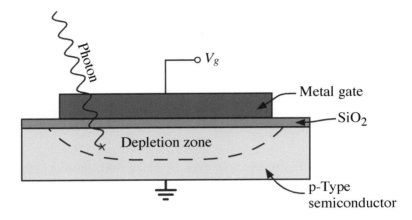

Figure 4.1: Cross section of a MOS capacitor used as a pixel in a CCD. A positive voltage V_g, usually about 10 volts, produces a depletion region in the semiconductor. An incident photon can pass through the thin metal gate and promote an electron in the valence band of the semiconductor into the conduction band. These photoelectrons collect under the metal gate. The MOS capacitor stores a number of electrons equal to the number of incident photons.

shows the structure of a MOS capacitor. The bottom layer is usually made of doped silicone to make it a p-Type semiconductor. A thin insulator, usually SiO_2, is deposited on the silicone substrate. A conducting gate is applied on top of the insulator. When a positive gate voltage V_g is applied, electron holes migrate away from the gate forming a **depletion region**—a region below the gate depleted of holes.[1] The electrons in the depletion region are trapped in the valence band so they are immobile. However, an incident photon with energy greater than the bandgap energy of the semiconductor can interact with a valence electron and

---

[1]To those acquainted with solid state physics the concepts of an electron hole, valence, and conduction bands, etc. will sound familiar. The detailed solid state physics isn't that important to our discussion, but if you want to learn more about the physics of semiconductors see Donald Neamen's book [17].

promote it to the conduction band. The newly freed electron will migrate toward the positively biased metal gate, but the insulator will trap it in the depletion region. The electron is effectively stored in the pixel. The electrons generated by interaction with photons are called **photoelectrons**.

Problem 4.1
The bandgap energy of Si is approximately 1 eV. How does this energy compare to energy of photons in the visible part of the spectrum?

If enough photoelectrons accumulate in the pixel they will begin to shield the positive voltage of the gate from newly generated photoelectrons and the pixel won't be able to hold any more charge. The maximum number of electrons that can be stored in a pixel is called the **full-well capacity**. The full well capacity is typically between 10,000 and 500,000 electrons. The number of electrons is directly proportional to the number of incident photons so CCDs have a linear response until the number of electrons in the pixel approaches the full well capacity. Figure 4.2 illustrates the response of a typical CCD. The device begins to depart from linearity and the device **saturates** as the number of electrons approaches the full-well capacity.

4.2   CCD READOUT

In order to take a picture with a CCD, we simply place it at the focal plane of a telescope behind a mechanical shutter. We clear all the charge out of the pixels and then open the shutter for the desired exposure time. When the exposure is done we have stored an electronic image. All we have to do now is to measure the charge in each pixel. Figure 4.3 shows how this is done. Each column of pixels is called a **parallel register**. The first step in reading out the charge in the device is to shift all of the charge in the parallel registers up by one row. The top row is shifted into a special row, called the **serial register**. The serial register isn't exposed to light so it is empty before the shift. Next, the charge in the serial register is shifted to the right. The charge in the right-most pixel is dumped onto a capacitor. This capacitor is just a special MOS capacitor and is sometimes called the **output node**. The voltage across the output capacitor is then amplified and converted to

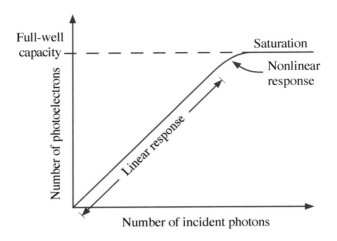

Figure 4.2: CCDs typically have a linear response with one photoelectron generated per photon until the pixel saturates as the MOS capacitor approaches full-well capacity.

a number by an analog-to-digital converter (ADC). The output of the ADC is stored in a computer. The charge is dumped from the output node and the charge in the serial register is again shifted to the right with the right-most pixels charge again dumped onto the output node and measured. The process is repeated until all of the charge has been read out of the serial register, then the next row is shifted into the serial register and the process of reading the serial register is repeated. In the end we have an array of numbers stored in computer memory with each number proportional to the number of photons that were collected by each pixel during the exposure.

Exactly how the charge is moved along the parallel and serial registers depends on the type of CCD, but Figure 4.4 shows a diagram of a few pixels of a three-phase CCD. The charge in each pixel is confined in the horizontal direction by permanent barriers called **channel stops**. The movement of charge along the parallel registers is controlled by adjusting the voltage on the gates that span the width of the CCD. For example, if ϕ_2 is set to some positive potential greater than ϕ_1 and ϕ_3 (see Figure 4.5 at time t_0) photoelectrons will be trapped in the pixel. This is the configuration of gate voltages used while the shutter is open and the CCD is collecting photoelectrons in its pixels.

Figure 4.5 shows the sequence of steps by which charge is transferred from one pixel to the next. The top of the figure shows a cross section of a

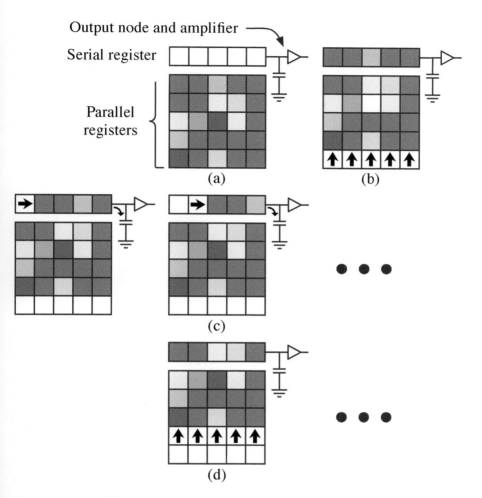

Figure 4.3: CCD readout process. (a) The CCD just after the shutter is closed. The number of charges in each pixel in the parallel register is proportional to the amount of light that was incident on the pixel. (b) The first step in the readout is to shift the charge in the parallel resisters up by one row. The charge in the top row is shifted into the serial register. (c) While the charge in the parallel registers is held in place, the charge in the serial register is shifted pixel by pixel onto the output node where the voltage is amplified and measured. (d) After the serial register has been read, the charge in the parallel registers is shifted up again by one row. The serial register is again read and the process repeated until the entire CCD's charge is readout and recorded.

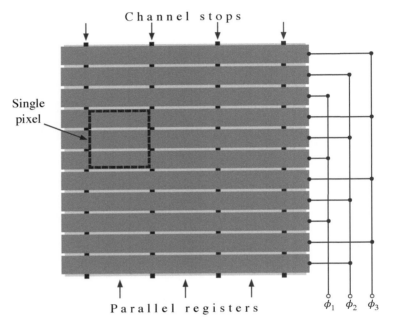

Figure 4.4: An enlarged diagram of a few pixels of a three-phase CCD showing channel stops that define the parallel registers and the gates that control the movement of charge along the parallel registers. During an exposure the gate voltages are set to a positive values with $\phi_2 > \phi_1 = \phi_3$ so that photoelectrons collect under the ϕ_2 gate.

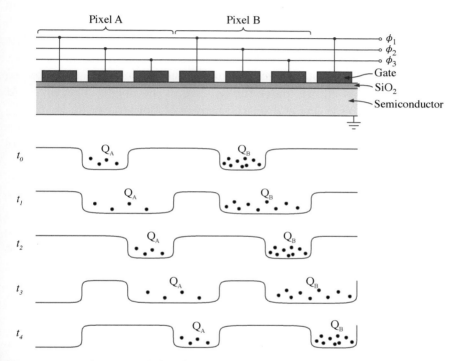

Figure 4.5: The top of the figure shows a cross section of a few pixels along one of the registers. The curves below the diagram show a sequence of electron potential energy curves used to transfer the charge to the right from one pixel to the next.

few pixels along one of the registers. The bottom of the figure shows the elections in the potential energy wells defined by setting the gate voltages ϕ_1, ϕ_2, and ϕ_3. At time t_0, $\phi_2 > \phi_1 = \phi_3$. This is how the gate voltages in the parallel registers are set during an exposure. Any photoelectrons generated in the depletion zone under a pixel in the parallel registers collects in the potential well under the ϕ_2 gate. Suppose a charge Q_A is trapped in pixel A and charge Q_B is trapped in pixel B. The process of shifting the charges to the right along the register starts at time t_1 when the readout electronics changes the potential on the ϕ_3 gates to match the potential ϕ_2 and the charge spreads out under the ϕ_2 and ϕ_3 gates. At time t_2, ϕ_2 has changed to force all the electrons under the ϕ_3 gate. At time t_3, the readout electronics change the potential ϕ_1 to match ϕ_3 and the electrons diffuse under the ϕ_1 gate of the neighboring pixel. At time t_4, the process has shifted the charge Q_A originally under the ϕ_2 gate of pixel A into pixel B under its ϕ_1 gate. As the readout electronics continue to cycle the gate voltages the charge moves from gate to gate along the register.

If the charge in one pixel exceeds the full-well capacity, the charge will leak over the potential barriers in the parallel register and into the adjacent pixels. This effect is called **blooming** shown in the image in Figure 4.6. The channel stops prevent charge from leaking along rows so the blooming only occurs along columns of pixels. Blooming ruins not only the signal in the saturated pixel, but data along entire columns of the CCD.

4.2.1 Charge Transfer Efficiency

Ideally *all* of the charge should be transferred from pixel to pixel during the readout. In practice this is impossible. The fraction of electrons that move from one pixel to the next during the transfer is called the **charge-transfer efficiency** (CTE). A CCD with a CTE $= 1$ is impossible, but modern CCDs must have CTEs very close to one. For a typical 1024 by 1024 pixel array, the pixel furthest from the output node must undergo 1024 transfers down the parallel register and then another 1024 transfers along the serial register. This means a total of $N = 2048$ transfers. The fraction of charge f delivered to the output node after N transfers is

$$f = \mathrm{CTE}^N.$$

To get 99% of the charge, $f = 0.99$, from the last pixel of a 1024 by 1024 CCD requires a CTE > 0.999995.

Figure 4.6: The two brightest stars in this image show obvious blooming. If you look closely you will see at least four of the fainter stars in the image are also blooming.

Problem 4.2

The DEIMOS camera on the Keck Telescope uses an array of 16 CCDs. Each CCD is 2048 by 4096 pixels. What CTE is required to transfer 99.9% of the charge from the last pixel to the output node of the CCDs?

In CCDs made from the MOS capacitors like the one shown in Figure 4.1, the photoelectrons collect and are transferred at the SiO_2–semiconductor interface. This type of CCD is called a **surface channel** CCD. Surface channel CCDs tend to have poor CTEs because charges get trapped at the interface. CCD designers avoid this problem by placing a p–n junction below the SiO_2 insulator. This shifts the depletion zone to the p–n junction and away from the trapping states at the SiO_2–semiconductor interface. CCDs using this design are called **buried-channel** devices. All modern CCD cameras used by astronomers employ buried-channel CCDs.

4.2.2 Analog to Digital Conversion

The charge on the output node of the CCD is amplified and then converted to a digital number by an ADC. The number from the ADC is a number of counts in **analog to digital units** (ADU) between 0 and the maximum digital output of the ADC. Most CCD cameras use 16-bit ADCs so the maximum count is $65{,}535 = (2^{16} - 1)$ ADU. The number from the ADC is proportional to the voltage from the amplifier and hence the number of electrons on the output node. The **gain** of the CCD is the number of electrons that correspond to one ADU.

The gain of a CCD is usually set so that the ADC reaches digital saturation before the pixels exceed their full-well capacity. This avoids the non-linear part of the response shown in Figure 4.2. The minimum voltage threshold of the ADC is set below the typical amplifier output voltage. This is done to make sure that noise added by the amplifier is sampled adequately. This offset means that even empty pixels register a few hundred ADU. This is **electronic bias** adds signal to the output and must be subtracted out to recover the original input signal (see Chapter 5).

Problem 4.3
Suppose a CCD camera uses a CCD with a full well capacity of about 100 000 electrons. The CCD is connected to a 16-bit ADC. What is the approximate gain of the of the system in electrons/ADU?

The read-out electronics actually measures the charge on the output node twice. First the electronics clear all the charge from the output node, and reads the signal level. Charge from the next pixel is then dumped onto the output node and read again. The **pixel value** stored in memory is the difference between these to readings. This process is called **correlated double sampling** and reduces the noise in the measurement. The process doesn't completely eliminate noise though. The typical **read noise** for state-of-the-art CCDs is only a few electrons. The read noise increases if the CCD is read out more rapidly. Most astronomical CCD cameras take at least 20 or 30 seconds to read out.

4.3 DARK CURRENT

At room temperature thermal agitation will occasionally kick electrons from the valence band to the condition band. This results in a steady trickle of charge into each pixel even if the CCD isn't being exposed to light. This flow of electrons is called **dark current**. A Si CCD can have a dark current of up to 10^4 electrons/second/pixel at room temperature, but it drops dramatically at lower temperatures. By treating the electrons as a free Fermi gas one can estimate the dark current as

$$\dot{n}_D = A T^{3/2} e^{-E_g/2kT}, \qquad (4.1)$$

where k is the Boltzmann constant, A is a constant that depends on the material, and E_g is the bandgap energy of the semiconductor [17]. Most research grade CCD cameras used at large observatories use liquid nitrogen (LN_2) to cool the CCD to about $-100°C$. This essentially eliminates dark current.

Problem 4.4
Assuming a dark current of 10^4 electrons/second/pixel at room temperature, estimate the dark current at $-100°C$.

The CCD in such a camera is sealed in a vacuum to insulate it from the atmosphere. The vacuum enclosure prevents frost from forming on the CCD. The camera housing must also have an insulated reservoir to hold the LN_2. This sort of camera housings is often referred to as a dewar.[2]

Less expensive CCD cameras use a thermoelectric cooler rather than LN_2. Thermoelectric coolers can typically cool the CCD by about 50°C below ambient temperature. This doesn't completely eliminate dark current, but does reduce it to a manageable level.

4.4 QUANTUM EFFICIENCY

The quantum efficiency of a CCD depends on wavelength. If the wavelength of the incoming light is too long the photons don't have enough energy to promote electrons from the valence band to the conduction band. Short wavelengths reflect off the surface, but photons with wavelengths in the visible band have a high probability of producing photoelectrons. Figure 4.7 shows quantum efficiency for two typical CCDs used by astronomers.

Most CCDs are **frontside illuminated** meaning that light enters silicon through the metal gates as seen shown in Figure 4.1. The metal in the gates is very thin so most of light passes right through the gate and into the depletion zone. These CCDs often have maximum quantum efficiencies of more than 60%. Frontside illuminated devices tend to have poor quantum efficiency on the blue end of the visible spectrum. **Backside illuminated** devices have much better blue sensitivity, but they must be made extremely thin (10–20 μm) so the light can pass through the silicone substrate into the depletion zone. Besides being expensive to manufacture, thinning causes problems in the near infrared part of the spectrum. The wavelength of infrared light is close enough to the thickness of the CCD that interference fringes form. The fringing depends on how the CCD is illuminated and is very difficult to correct for later while processing the images.

4.5 EXAMPLE CCD CAMERAS

The Table 4.1 below shows the specifications for two astronomical cameras. The STL-1001E is a commercial camera manufactured by SBIG Astronomical Instruments (https://www.sbig.com). The NASAcam

---

[2]A dewar is actually just a vacuum flask used to store cold or hot liquids. It was invented by Sir James Dewar in 1892.

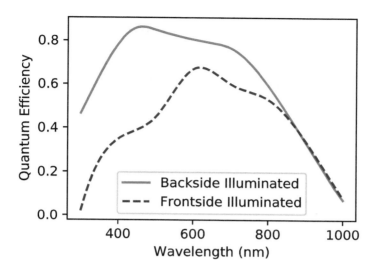

Figure 4.7: Quantum efficiency curves for a frontside illuminated and a backside illuminated CCD. Notice that the backside illuminated device has better quantum efficiency at short wavelengths.

is a custom-built camera for the 31-inch telescope at Lowell Observatory used by the National Undergraduate Research Observatory (http://www.nuro.nau.edu). The STL-1001E is thermoelectrically cooled to about 50°C below the ambient temperature, but even then the dark current is significant. The NASAcam operates at −110°C so its dark current is negligibly small.

SUPPLEMENTARY PROBLEMS

Problem 4.5 CCDs made of silicon can detect visible light, but the bandgap energy is too large to large to detect infrared light with wavelengths longer than about 1 μm. CCDs used for infrared imaging are often fabricated using indium antimonide (InSb). The bandgap energy of InSb is about 0.2 eV. What is the longest wavelength light that could be detected by an InSb CCD?

Problem 4.6 The average thermal energy of electrons in a solid is approximately equal to kT. How does this energy compare to the bandgap energy of Si at room temperature? Given this calculation how do you

Table 4.1: Specifications for two typical CCD cameras used by astronomers

	STL-1001E	NASAcam
CCD	Kodak KAF-1001E	Loral 2K×2K
Array size	1024 × 1024	2048 × 2048
Pixel size (μm)	24 × 24	13.5 × 13.5
Full-well capacity (e-)	150 000	125 000
Gain (e-/ADU)	2.2	1.8
Read noise (e-)	17	14
Operating Temp. (°C)	~ -30	-110
Dark current (e-/s/pixel)	34 at 0°C	~ 0 at -110°C

explain the fact that room temperature CCDs have significant dark current?

Problem 4.7 CCDs cooled with LN_2 are typically cooled to -100°C, but thermoelectric coolers can only cool them to about 50°C below ambient temperature. Assuming a dark current of 10^4 electrons/second/pixel at room temperature, estimate the dark current in a thermoelectrically cooled CCD if the ambient temperature is 20°C.

Problem 4.8 Compare the dark current in a Si CCD ($E_g \approx 1$ eV) and an InSb CCD ($E_g \approx 0.2$ eV) at the typical operating temperatures of an LN_2 cooled CCD. By how much would the dark current be reduced if it were cooled with liquid He? The boiling point of liquid He is 4.2 K at one atmosphere.

CHAPTER 5

Image Processing

The data from a CCD detector is recorded as a two dimensional array of pixel values. Ideally a digital image would be a perfect record of the light coming from some astronomical object. Unfortunately, raw CCD images are far from perfect. They are affected by dark current, noise in the readout electronics, quantum efficiency that varies across the CCD, and even dust on the optics. Fortunately, once the pixel values are stored in a computer we can use **image processing** software to manipulate the digital image to correct most of these defects. The companion website[1] contains a list of some commonly used image processing packages used by astronomers. You will need access to one of these packages in order to do most of the problems in this chapter. You can also download the images needed to do the problems from the text's companion website.

Before we can begin to process the images we need to see them. The first part of this chapter discusses how we can display digital images and modify the display to bring out subtle features that might not be visible otherwise. The next sections describes how we use image processing software to remove defects in raw CCD images. Finally, we explore ways to combine images in order to improve the signal-to-noise ratio, create color images, or create an image of a large patch of the sky.

5.1 DISPLAYING IMAGES

The pixel values that make up a digital image from a CCD are proportional to the number of electrons stored in each pixel. In order to create a picture from the pixel-value array, an image display program must **map** a pixel value to a grayscale level or a color. An image displayed using a

---

[1]`https://mshaneburns.github.io/ObsAstro/`

DOI: 10.1201/9781003203919-5

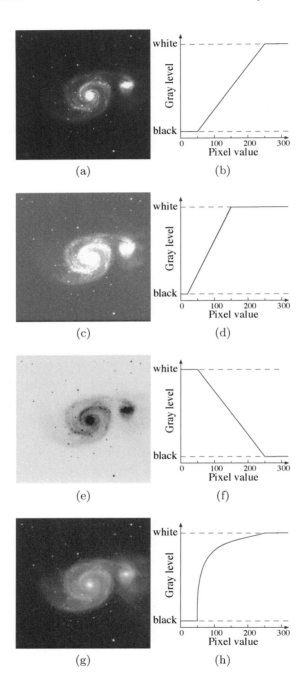

Figure 5.1: An R-band image of M51 with different grayscale mappings. The corresponding grayscale map is shown to the right of each image.

color mapping is a **false-color image**, since each color represents a pixel value rather than the true color of the object. **Grayscale images** assign levels of gray to pixel values.

Figure 5.1(a) shows an image of the galaxy M51. Figure 5.1(b) shows the grayscale mapping of pixel values to shades of gray. In this image pixel values greater than 250 are represented as white and values below 50 are black with a linear scaling of gray levels between 50 and 250. Figure 5.1(c) shows the same digital image with a different grayscale mapping—pixel values below 20 are black and those above 150 are black. This mapping brings out more subtle detail in the spiral arms, but loses detail in the galaxy's center. Figure 5.1(d) shows the corresponding grayscale mapping. This mapping has a larger slope which produces more contrast in the faint spiral arms. Figure 5.1(e) shows an **inverted grayscale** map where low pixel values are mapped to white and high pixel values are mapped to black producing a **negative image**. Negative images allow you to see faint detail that isn't always apparent in positive images. We can use nonlinear mapping to bring out faint details, but keep detail in the bright part of the image. Figure 5.1(g) and (h) show an image with a nonlinear grayscale map. In this case the grayscale is computed from the logarithm of the pixel values. This mapping has a large slope for small pixels values allowing you to see faint details in the spiral arms, but has a small slope for large pixel values so that you don't lose detail in the bright galaxy center.

Since your eyes are better at distinguishing colors than disguising levels of gray, a color mapping will allow you to see subtle features not apparent in grayscale images. Figure 5.2 shows a false-color image of the same galaxy as those shown in Figure 5.1. The color mapping is shown to the right of the image. The faint halo outside of stars surrounding both M51 and it's companion galaxy is easily visible in the false-color image, but difficult to see in the grayscale images.

Problem 5.1

Download the image files from the text's companion website. Use one of the image display software packages listed listed on the site to display the image of M 51 (`M51_R_p.fits`).

(a) Adjust the grayscale to make the galaxy arms clearly visible.

(b) Invert the grayscale to create a negative image.

(c) Display the image with a false-color color map.

Figure 5.2: False color image of the same digital image shown in grayscale images in Figure 5.1. The vertical bar on the right shows the color mapping to pixel values.

The human eye has three different kinds of color receptors called cones each having a different color sensitivity. One cone is most sensitive to blue light, one is most sensitive to green, and the other is most sensitive to red. Computer monitors display any color that can be seen by humans by adding together red, green, and blue light. This means that we can create a so-called true-color image of an astronomical image by combining three digital images—one taken using a red filter, one using a green filter, and one using a blue filter. Images produced in this way are called **RGB images**. Figure 5.3 is an RGB true-color image formed by assigning an image taken with a B-band filter to blue, one with a V-band filter to green, and one with an R-band filter to red. This approximates what might be seen by a human eye, if it were sensitive enough. Of course, the color assignment is arbitrary. One could also assign the V-band to blue, the R-band to green, and the I-band to red. This would be a false-color image, but we could see how the infrared emission compares to the other two bands.

Figure 5.3: RGB true-color image of M51 produced by using B, V, and R-band images.

5.2   IMAGE ARITHMETIC

A digital image is just an array of numbers. Each pixel value depends on not only the number of photons that strike the pixel, but also things like dark current, electronic offset, and CCD noise. We can modify the original image to more closely represent the photon flux by modifying each pixel value in the array.

Before we discuss how to correct an image, let's develop some notation and conventions so we can represent doing mathematical operations on digital images. Suppose we have a digital image that is N_x by N_y pixels in size. The image is just an array of pixel values S_{xy} where $1 \leq x \leq N_x$ and $1 \leq y \leq N_y$. The complete image is just a giant matrix $\mathbf{S}$. We do can do image arithmetic element by element. Adding two images $\mathbf{A}$ and $\mathbf{B}$ yields another image $\mathbf{C}$ and is represented as

$$\mathbf{C} = \mathbf{A} + \mathbf{B}$$

and means that $C_{xy} = A_{xy} + B_{xy}$ for all $1 \leq x \leq N_x$ and $1 \leq y \leq N_y$. This is just how we would treat an ordinary matrix, i.e., like ordinary matrices we can subtract two images or add, subtract, multiply and divide images by a constant. The only difference between image arithmetic and matrix arithmetic is multiplication and division. When working with digital images these operations proceed element by element. If $\mathbf{C}$ is the product of $\mathbf{A}$ and $\mathbf{B}$ then

$$\mathbf{C} = \mathbf{A} * \mathbf{B}$$

means $C_{xy} = A_{xy}B_{xy}$ for all $1 \leq x \leq N_x$ and $1 \leq y \leq N_y$. We use the '$*$' operator to distinguish image multiplication from matrix multiplication. We define image division in the same way,

$$\mathbf{C} = \mathbf{A}/\mathbf{B}$$

means $C_{xy} = A_{xy}/B_{xy}$ for all $1 \leq x \leq N_x$ and $1 \leq y \leq N_y$.[2]

We can go further to define any mathematical operation on an image to be done element by element:

$$\mathbf{C} = \log(\mathbf{A}) * \mathbf{B}$$

gives $C_{xy} = \log(A_{xy})B_{xy}$ for all $1 \leq x \leq N_x$ and $1 \leq y \leq N_y$. We can also define operations on groups of images. For example,

$$\mathbf{S} = \text{median}(\mathbf{A_1}, \mathbf{A_2}, \mathbf{A_3}, \mathbf{A_4}) = \text{median}(\mathbf{A_i})$$

---

[2] We can use the '/' symbol for image division without confusion since there's no conventional definition for matrix division.

gives the image **S** in which in each pixel value is the median of the values of the corresponding pixels in $\mathbf{A_1}, \mathbf{A_2}, \mathbf{A_3}$, and $\mathbf{A_4}$.

Problem 5.2

The FITS files `gal_ref_p.fits` and `gal_new_p.fits` are two artificial images of an elliptical galaxy. Write a program or use some image processing software to subtract `gal_ref_p.fits` from `gal_new_p.fits` and display the result. Do you see anything unusual in the difference image?

Problem 5.3

Images `stars01_p.fits`, `stars02_p.fits`, and `stars03_p.fits` are three images of a cluster of five stars. Write a program or use some image processing software to take the mean of these three images. Display the result. How does it compare to the original images? Note that there is a slight misalignment between these three images. We'll learn how to align images later in the chapter.

5.3 CCD DATA CORRECTION

Below saturation CCDs have a linear response. That means that if **R** is the raw digital image from CCD, then

$$\mathbf{R} = \mathbf{E} * \mathbf{S} + \mathbf{B}, \tag{5.1}$$

where **S** is the underlying perfect image. You can think of **E** as a efficiency factor for each pixel. **E** is a *multiplicative* factor for each pixel that accounts for pixel to pixel variations in the response of the CCD. It accounts for pixel-to-pixel quantum efficiency variations, dust or fingerprints on the optics or filters, and optical vignetting.[3] You can think of **E** as a collection of pixel efficiency factors. The background image **B** is signal *added* to the perfect image by electronic bias, dark current, cosmic rays, and any other signal that doesn't come from an astronomical source.

We can recover the original image by doing a little image arithmetic

---

[3]Optical vignetting is a reduction of an image's brightness at the edge compared to the image center. It is often cause by misaligned filters or optical elements.

on the output image,

$$\mathbf{S} = \frac{\mathbf{R} - \mathbf{B}}{\mathbf{E}}. \tag{5.2}$$

Of course, this assumes we have some way to estimate the background image, $\mathbf{B}$ and the efficiency image $\mathbf{E}$.

5.3.1 Background Image

There are three primary sources of background signal: electronic bias, dark current, and cosmic rays. Cosmic rays are high energy particles that "kick" electrons into the conduction band as they pass through the CCD. Cosmic rays only affect at most a few pixels and aren't generally a problem. One can eliminate them by averaging the signal over pixels surrounding the cosmic ray hit, but this is only a cosmetic correction. Any data affected by cosmic rays is lost. Cosmic rays can't be corrected by subtracting a background image. We can, however, estimate the electronic bias and dark current contributions to the background image.

5.3.1.1 Bias Frames

Electronic bias is purposely added to images so that the ADC functions properly. We can estimate the bias by clearing the CCD and immediately reading again with zero exposure time. In this case the input signal $\mathbf{S} = 0$ and the dark current signal are both zero. Frames taken in this way are called **bias frames** or **zero frames**. There is noise in each bias frame so we typically take about ten bias frames and average them together to get a master bias image. If the CCD's dark current is negligible then the master bias frame a good estimate of the background.

Some CCDs cameras estimate the bias by **overscanning** the horizontal register. The CCD electronics overscan the horizontal register by continuing to read the output gate after the last pixel of charge has been read. Overscanning results in an image that has more pixels in the horizontal direction than the physical size of the image. The overscanned pixels are effectively a mini bias image. If you are processing images from an overscanned CCD, you can use the overscanned pixels to estimate the bias then trim the overscanned region from each image. To be safe you should always create a master bias frame even if you are using an overscanning CCD.

Problem 5.4

The ten images `bias01.fits` through `bias10.fits` are raw bias frames. Create a master bias image by averaging these ten images.

5.3.1.2 Dark Frames

Dark current causes charge to build in each pixel even if it isn't exposed to light. Dark current varies from pixel to pixel. We can estimate the dark current by taking a **dark frame**, an image produced by clearing the CCD then waiting a period of time before reading the CCD. We are essentially taking an image without opening the shutter. If dark current is a problem we usually take dark frame with an "exposure time" equal to or longer than the images we plan to take. The amount of charge built up is governed by the Poisson distribution so there is some noise in every dark frame. We reduce the noise by creating a master dark current image $\mathbf{D}$ by taking a set of N dark frames $\mathbf{d_i}$, subtracting the bias, dividing each by the corresponding dark exposure time $t_{d,i}$, and then averaging or taking the median of the images:

$$\mathbf{D} = \text{median} \left(\frac{\mathbf{d_1} - \mathbf{Z}}{t_{d,1}}, \frac{\mathbf{d_2} - \mathbf{Z}}{t_{d,2}}, \dots \frac{\mathbf{d_N} - \mathbf{Z}}{t_{d,N}} \right), \tag{5.3}$$

where $\mathbf{Z}$ is the master bias frame.

Assuming that thermal charge builds linearly with time, we can estimate the dark current contribution to the background in any image by multiplying $\mathbf{D}$ by that image's exposure time t. The total background image is then

$$\mathbf{B} = \mathbf{D}t + \mathbf{Z}. \tag{5.4}$$

Most research grade CCD cameras are cooled to a low enough temperature ($\sim 100°$ C) that dark current becomes insignificant. If you have data taken with one of these cameras you don't need to correct for dark current.

5.3.2 Efficiency Images

We estimate the efficiency image by exposing the CCD to something that is uniformly illuminated. Uniform illumination means that the pixel values in $\mathbf{S}$ are all the same. Uniformly illuminated images are called **flat-field images** or more simply **flats**. Using equation (5.1) we see that

$$\mathbf{E} = (\mathbf{R} - \mathbf{B})/S, \tag{5.5}$$

where S is the pixel value of all the pixels in $\mathbf{S}$. We can get $\mathbf{B}$ from the process outlined in the last section. $\mathbf{R}$ is the output image, but we don't generally know S. Not knowing S isn't crucial to getting the relative efficiencies of the pixels. We can set S to be almost anything and the resulting image will be proportional to $\mathbf{E}$. One way to estimate S is to set it equal to the average pixel value in $\mathbf{R} - \mathbf{B}$,

$$S = \bar{S} \equiv \sum_{x=1}^{N_x} \sum_{y=1}^{N_y} \frac{R_{xy} - B_{xy}}{N_x N_y}.$$

In practice, we typically set S equal to the *median* pixel value $\tilde{S}$ of $\mathbf{R} - \mathbf{B}$. The median is less sensitive to outliers that might be produced by cosmic rays or bad pixels. Setting S as either the mean or median pixel value ensures that the typical pixel values in the resulting efficiency image are close to one.

The quantum efficiency of a CCD is wavelength dependent and varies from pixel to pixel so you must make an efficiency image for each filter. Astronomers usually take about five flat-field images in each filter and combine them to reduce the noise in the final efficiency map. How they combine the images depends on how the images were taken.

5.3.2.1 *Twilight flats*

By far the most common way to create flat-field images is to take images of the twilight sky. The twilight sky isn't perfectly uniform, but over the small field of view of most telescopes it is adequately uniform near the zenith. One problem with this technique is that the sun is either setting or rising while you are taking images. This means that you must scale the images by their median pixel-value before you combine them. Another problem is that each flat probably has a different exposure time so each image will collect a different of dark-current charge.

The image processing for flats is getting complicated so let's outline an algorithm. We'll assume we have already created a dark current frame $\mathbf{D}$ and a master bias frame $\mathbf{Z}$. Suppose we have N raw flat-field frames $\mathbf{f_i}$ with the corresponding exposure times t_i.

1. Subtract the background from each raw flat to get the background-subtracted flats $\mathbf{F_i}$,

$$\mathbf{F_i} = \mathbf{f_i} - \mathbf{D}t_i - \mathbf{Z}.$$

2. Compute the median pixel value $\widetilde{F}_i$ of each flat field image $\mathbf{F_i}$.

3. Compute the efficiency frame or the master flat-field image by taking the median of the normalized flats,

$$\mathbf{E} = \text{median} \left(\frac{\mathbf{F_1}}{\widetilde{F}_1}, \frac{\mathbf{F_2}}{\widetilde{F}_2}, ... \frac{\mathbf{F_N}}{\widetilde{F}_N} \right).$$

Problem 5.5

The images `flat01.fits` through `flat05.fits` are flat field images. Use them to produce a single efficiency image. Don't forget to subtract the background using the master bias image created in Problem 5.4. You may assume that the dark current is negligible.

5.3.2.2 Dark-sky flats

Except for individual stars, the night sky is perfectly uniform directly overhead, but doesn't emit much light. If the exposure time is long enough to get a significant signal from the sky the image will also have stars. One way around this problem is to take a large number of deep sky images, but shift the telescope between each image. When you take the median of all these images, the bright pixels containing stars won't contaminate the final efficiency map.

Because of the long exposure times required to get good dark-sky flats this technique is really only useful if your observing program already includes taking long time exposures of different regions of the sky. If you place the object or objects of interest in different locations on the image you can take the median of these images to create the efficiency map. You can use the same processing algorithm for processing dark-sky flats as outlined for twilight flats except you need to include many more images to eliminate the effects of stars and get enough signal to eliminate noise in the final efficiency frame.

5.3.2.3 Dome flats

Another way to produced a flat field is to point the telescope at a uniformly illuminated screen inside the dome. One advantage of this technique is that the astronomer controls the illumination so all of the images taken with the same exposure time should yield the same signal level. You can take dome flats during the day. One problem with dome flats is that it is very difficult to uniformly illuminate the screen. If the weather is good, twilight sky flats are usually a better option.

5.3.3 Processing Image Data

Once you have created all of the calibration images—the master bias $\mathbf{Z}$, dark current $\mathbf{D}$, and efficiency $\mathbf{E}$ frames—you can correct raw CCD images $\mathbf{R_i}$ to produce corrected images $\mathbf{S_i}$,

$$\mathbf{S_i} = \frac{\mathbf{R_i} - \mathbf{D}t_i - \mathbf{Z}}{\mathbf{E}}, \tag{5.6}$$

where t_i is the exposure time for $\mathbf{R_i}$. If you then multiply $\mathbf{S_i}$ by the CCD gain you should have an image where the pixel values are equal to the number of photoelectrons.

Figure 5.4 shows a master bias frame (a), an R-band flat-field image (b), a raw galaxy image (c) and a final image (d) corrected using equation (5.6). The image was taken with a CCD camera with negligible dark current so $\mathbf{D} = 0$. The final image doesn't look much different than the original image, but the actual pixel values have changed. In order to get accurate photometry you will need to use the final processed image.

Problem 5.6

The image `stars01.fits` is a raw CCD image of a star field. Use the background image created in Problem 5.4 and the flat field efficiency image created in Problem 5.5 to correct this image. You may again assume that the dark current is negligible.

Suppose you have just completed a night's observing run, taken a well deserved rest, and now you want to process your data. You took a set of bias frames $\mathbf{z_1}, \mathbf{z_2}, ..., \mathbf{z_{N_z}}$, and a set of dark frames $\mathbf{d_1}, \mathbf{d_2}, ..., \mathbf{d_{N_d}}$. You needed data in the B-band and V-band so you took twilight flats using both filters, $\mathbf{f_1^B}, \mathbf{f_2^B}, ..., \mathbf{f_{N_f}^B}$ and $\mathbf{f_1^V}, \mathbf{f_2^V}, ..., \mathbf{f_{N_f}^V}$. Let's sumarize how you might go about processing your raw data files taken with B filter $\mathbf{R_1^B}, \mathbf{R_2^B}, ..., \mathbf{R_{N_B}^B}$ and the V filter $\mathbf{R_1^V}, \mathbf{R_2^V}, ..., \mathbf{R_{N_V}^V}$.

1. Create a master bias frame $\mathbf{Z}$,

$$\mathbf{Z} = \text{average}\left(\mathbf{z_1}, \mathbf{z_2}, ..., \mathbf{z_{N_z}}\right).$$

2. Create a dark current frame $\mathbf{D}$,

$$\mathbf{D} = \text{median}\left(\frac{\mathbf{d_1} - \mathbf{Z}}{t_{d,1}}, \frac{\mathbf{d_2} - \mathbf{Z}}{t_{d,2}}, ..., \frac{\mathbf{d_N} - \mathbf{Z}}{t_{d,N}}\right).$$

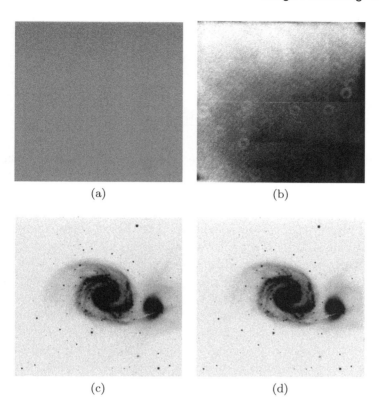

Figure 5.4: Images used to correct an image taken with a CCD. The dark current for this camera is negligible so no dark frame was needed. (a) Master bias frame. (b) R-band flat-field. (c) Raw R-band image of the galaxy M51. (d) Corrected galaxy image.

3. Create a master flat-field frame for each filter,

$$\mathbf{E^B} = \text{median} \left(\frac{\mathbf{F_1^B}}{\widetilde{F_1^B}}, \frac{\mathbf{F_2^B}}{\widetilde{F_2^B}}, ..., \frac{\mathbf{F_{N_f}^B}}{\widetilde{F_{N_f}^B}} \right),$$

and

$$\mathbf{E^V} = \text{median} \left(\frac{\mathbf{F_1^V}}{\widetilde{F_1^V}}, \frac{\mathbf{F_2^V}}{\widetilde{F_2^V}}, ..., \frac{\mathbf{F_{N_f}^V}}{\widetilde{F_{N_f}^V}} \right),$$

where $\mathbf{F_i^B} = \mathbf{f_i^B} - \mathbf{D}\, t_{f,i}^B - \mathbf{Z}$, and $\mathbf{F_i^V} = \mathbf{f_i^V} - \mathbf{D}\, t_{f,i}^B - \mathbf{Z}$. The times $t_{f,i}^B$ and $t_{f,i}^V$ are the flat-field exposure times.

4. Process the raw images using the master flats to get the final corrected images,

$$S_i^B = \frac{R_i^B - D\,t_i^B - Z}{E^B}$$

and

$$S_i^V = \frac{R_i^V - D\,t_i^V - Z}{E^V}.$$

Sometimes you might continue the processing to eliminate cosmic ray hits, or bad pixels in the array. For some detectors, not typically CCDs, you might have to correct the data for the nonlinear gain of the device. Some CCD cameras produce images with an overscan region in which case you would have to add another step to correct for the overscan and trim the images. For CCD cameras that have negligible dark current you can eliminate the dark current processing steps.

You could write your own programs to do all the image processing and some astronomers do, but usually they use an image processing program that is designed specifically for dealing with astronomical images. Astronomical image processing programs like IRAF will process batches of images to create master bias frames, dark current frames, and efficiency maps using a few simple commands.[4] Sophisticated image processing programs are quite flexible and will allow you to process the images any way you want. For example, if you found that the raw bias frames were contaminated by cosmic ray hits you might choose to take the median of the images rather than the average to create the master bias frame.

Problem 5.7
Use the master bias image of Problem 5.4 and the flat field efficiency image created in Problem 5.5 to correct the images `stars01.fits` through `stars05.fits`, `gal_ref.fits`, and `gal_new.fits`. You may want to write a program that reads in a list of file names and corrects all the images in the list. This will be handy later on when you need to process a large number of images. Dark current is negligible.

---

[4]If you configure IRAF and the image files properly, then you can do the four processing steps outlined above with just four commands.

5.4   COMBINING IMAGES

Suppose you are trying to measure the brightness of a very faint star. You would need a very long exposure time to get a good signal-to-noise ratio. There are a number of problems with using one very long exposure time. For example, the telescope you are using will probably track very well for 15 minutes, but over a few hours a small tracking error could ruin the image. The dark current might be negligible for short exposure times, but might saturate the CCD for long exposures. Longer exposures also mean a greater chance that a cosmic ray might pass through the middle of your star and ruin the entire exposure. You can avoid all of these problems by taking a series of short exposures and combining them. You do have to make sure each individual exposure time is long enough that CCD read noise doesn't make a significant contribution. You can eliminate tracking errors by aligning the images before adding them. You can throw out images in which the data was ruined by cosmic rays[5].

5.4.1   Aligning Images

Before you can combine images you must align them. This consists of two steps: finding the offset between each image and somehow shifting the images. You can accomplish both these steps in a variety of ways and the whole process can be quite complicated. We'll explore two of the simplest algorithms. You can find descriptions of more sophisticated processing techniques in Chromey[7] or Martinez and Klotz[15].

If there are some relatively bright stars in our images we can compute the **centroids** of the stars. This gives the location of the center of the star to a fraction of a pixel. The centroid is computed a little like the center of mass of some object, but in this case mass is replaced by the pixel value. The first step in this algorithm is to isolate the star, that is, determine which pixels in the image contribute to the star's image. One technique is to use all of the pixels in a square region centered on the star's brightest pixel. Set the size of the region's sides equal to twice the FWHM of the stellar profile. You can then compute the horizontal and

---

[5]Most astronomical image processing software packages include some very sophisticated algorithms for handling cosmic rays. One example is Astro-SCRAPPY (https://github.com/astropy/astroscrappy). It is an Astropy (https://www.astropy.org/index.html) affiliated cosmic ray package.

vertical centroid position from the sums,

$$x_c = \frac{\sum_x \sum_y x(S_{xy} - B)}{\sum_x \sum_y (S_{xy} - B)}, \quad y_c = \frac{\sum_x \sum_y y(S_{xy} - B)}{\sum_x \sum_y (S_{xy} - B)}, \quad (5.7)$$

where the sums are only over pixels in the square region and B is background value computed from some region outside the square.

Problem 5.8
Download the image `stars06_p.fits` from the companion website. Write your own program or use an image processing package to find the centroids of the stars in the image. Assume the background is 275. Take the bottom left pixel to have coordinates $x = 1$ and $y = 1$.

Typically there are many stars in each image. You can determine an offset between the two images by centroiding all of the available stars and averaging the offsets between each stars coordinates. If there are only extended objects like large galaxies in the image we would have to use a more sophisticated technique.

Computing the centroid is simple, but isn't particularly robust. The algorithm will give poor results if the background is large compared to the brightness of the star, if field is crowded, or if light from an extended object contaminates the star's signal. Most sophisticated image processing programs can fit a function to a star's PSF. Fitting all of the stars in the image to a theoretical PSF will give good positions even in crowed fields. The disadvantage of PSF fitting is that the algorithms are often quite complicated.

Once the we have determined the positions of the centers of the stars, we can determine by how much each images is offset from the others. We can take the star positions in one of the images as the reference positions. We then compute the offsets by subtracting the coordinates of the stars in reference image from the coordinates of the corresponding stars in the rest of the images.

Problem 5.9
Use the program you used to solve Problem 5.8 or use some image

processing software to compute the positions of the stars in the images stars06_p.fits through stars10_p.fits. Use these images to determine the offsets between the images. Use stars06_p.fits as the reference image.

Once the offsets are known the simplest image shifting algorithms proceed in three steps. The first step is to align images to within one pixel by simply relabeling the pixel indices. The software then does a fractional pixel shift by interpolating a pixel value from the surrounding pixels. Finally, the software trims all the images to the same size. This step eliminates all the regions in each image that don't overlap. Once you have aligned the images you can often combine them by simply adding them together. The resulting image will have a larger SNR than any one of the individual images.

Sometimes shifting the images won't align the images well enough to combine. Suppose for example that the CCD camera wasn't taken off and then remounted on the telescope. If it was mounted in a slightly rotated position, the images will be rotated too. Cameras with a very wide field of view often produce images with some optical distortion. In this case, each pixel in the image covers a slightly different angular patch on the sky. If you want to combine data from two different cameras, they might have a different plate scale. In all these cases you will have to use a far more sophisticated algorithm than the one described above. One very useful method is called variable-pixel linear reconstruction [10]. This technique, sometimes called Drizzle, was developed for processing images for the Hubble Deep Field. Drizzle preserves photometry and can actually enhance the resolution of the final image if enough individual images are combined.

Besides improving the SNR or resolution of an image, might want to use several images taken through different filters to produce a color image or create a mosaic image to capture a bigger angular size than your camera's field of view. These algorithms can be quite tricky to code. Some of the software packages that allow this sort of specialized processing are listed on the companion website[6].

---

[6]https://mshaneburns.github.io/ObsAstro/

SUPPLEMENTARY PROBLEMS

Problem 5.10 Download a FITS format image of your choice from the ESA/Hubble Space Telescope's *Datasets for education and for fun* website.[7] Display the downloaded image with different grayscale settings and color maps.

---

[7]https://esahubble.org/projects/fits_liberator/datasets/

CHAPTER 6

Photometry

The science of measuring the intensity of light is called photometry. The first astronomers to quantify the brightness of stars used only their eyes to estimate brightness. A modern photometric measurement of a celestial object is a quantification of flux or photon count rate. In Chapter 3, we described how an object's magnitude is related to flux and photon count rate. We also saw that the measurement depends on the characteristics of the telescope, the camera, and even the atmosphere. Astronomers have devised standard photometric systems in order to characterize these effects. Astronomers can compare measurements from different telescopes and instruments by using filters to replicate the response functions of standard systems. This chapter describes some of the most commonly used standard systems. We will then explore techniques for extracting photon count rates from digitized images. We will also learn how to correct for atmospheric extinction. Finally we will learn some techniques for transforming measurements taken using a particular instrument to a standard photometric system.

6.1 STANDARD PHOTOMETRIC SYSTEMS

There are more than a hundred different photometric systems in use today. Many of these systems were devised for specific purposes. Some have wide passbands (30 nm or wider) and some are quite narrow for measuring particular spectral lines. In Chapter 3, we discussed one of the most widely used systems, the Johnson-Cousins $UBVRI$ system. Another commonly used systems is the $u'g'r'i'z'$ system developed for the Sloan Digital Sky Survey and is discussed below. See Bessell [1] for a review of several commonly used photometric systems.

DOI: 10.1201/9781003203919-6

6.1.1 The Johnson–Cousins $UBVRI$ System

The first well-characterized standard system was established by Johnson and Morgan in 1953 [13]. The system consisted of just three passbands, U, B, and V. The system was later extended into the red (R) and infrared (I) parts of the spectrum. Table 3.1 is reproduced as Table 6.1 below. It lists the central wavelength (λ_0), approximate filter bandwidth ($\Delta\lambda$) and the approximate flux (F_λ) for a zero magnitude star for all the filters in the $UBVRI$ System.

Table 6.1: Johnson–Cousins $UBVRI$ system response function effective wavelengths and widths. (Data from Bessell [1].)

Bandpass	U	B	V	R	I
λ_{eff} (nm)	366	436	545	641	798
$\Delta\lambda$ (nm)	65	89	84	158	154

The $UBVRI$ photometric system was developed before CCDs were used for photometry. The system was originally developed using photomultiplier detectors and specific filters. If we want to make measurements with a CCD, we have to make sure that the response of our CCD camera mimics the Johnson–Cousins system or that we're at least able to transform our magnitudes to standard $UBVRI$ magnitudes. This can be difficult since the passbands were defined by a combination of the sensitivity of the photomultiplier, the filters, and for the U filter the atmosphere.

Most CCDs are red sensitive and have poor U response so most CCD observations are done with B, V, R, and I filters. Figure 6.1 shows the normalized response curves for the $UBVRI$ system. To use the $BVRI$ system, the CCD/filter combination has to be chosen to match the Johnson-Cousins photomultiplier/filter systems as closely as possible.

In order to calibrate our system, we need a set of standard stars for which the colors and magnitudes have already been measured. Originally 290 stars defined the UBV system. Most of these stars are brighter than sixth magnitude. These are too bright for most CCD/telescope combinations. Since then a number of observers have carefully calibrated dimmer stars specifically for use with CCD cameras.

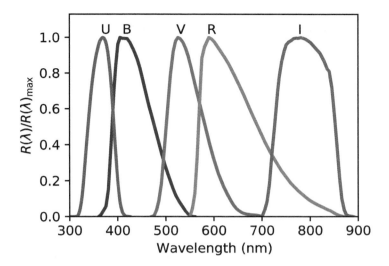

Figure 6.1: Normalized response $R(\lambda)$ for the $UBVRI$ standard photometric system.

6.1.2 Magnitude Zero Points

Catalogs of standard stars list the magnitudes and colors of all standard stars. However, there is some arbitrariness in the magnitude system. In order to completely specify the magnitudes in a particular photometric system, we point we have to define the **zero-point** of the system. That is, we have to define the flux of a star that has zero magnitude in the system. Tying fluxes to magnitudes turns out to be quite complicated, but one way around this problem is to define some star as the primary standard star and assign it a magnitude and color indices of zero. This is precisely what was done in the 1950s with the UBV system. The chosen star was Vega. It was chosen to have $V = 0$ as well as $B - V = 0$ etc. Vega is an A0 star, so other A0 stars scattered around the sky were chosen as secondary standards. Vega probably wasn't the best choice—it's a variable star and has excess IR radiation due to a circumstellar dust shell or ring. The modern zero-point definitions of the UBV system are based on a combination of modeling and spectrophotometric observations. In this modern system Vega has $V = 0.03$.

One difficulty with a zero point based on a particular class of star is that the zero point then depends on the somewhat complicated spectrum of the zero point star. This complicates comparing the magnitudes of stars with vastly different spectra. One way around this problem is to define the zero-point spectrum in a different way. One such zero point system is the AB magnitude system. The magnitudes in this system are defined such that a monochromatic flux (flux per unit frequency) of 3.63×10^{-20} erg sec^{-1} cm^{-2} Hz^{-1} gives zero magnitude.[1] This means that if the monochromatic flux, F_ν, is measured in erg sec^{-1} cm^{-2} Hz^{-1} then

$$m_{AB}(\nu) = -2.5 \log(F_\nu) - 48.60. \qquad (6.1)$$

In this system an object with constant F_ν has a color index of zero. One can convert AB magnitudes to other systems, but the magnitudes in any system are dependent on the assumed bandpass and source's spectrum, so the conversions can be quite difficult to compute. Table 6.2 shows the differences between $UBVRI$ magnitudes using the Vega and AB magnitude zero-points calculated by Blanton and Roweis [4].

Table 6.2: Difference between Vega and AB magnitudes zero-points for the $UBVRI$ bandpasses computed by Blanton and Roweis [4].

Bandpass	$m_{AB} - m_{Vega}$
U	0.79
B	−0.09
V	0.02
R	0.21
I	0.45

Problem 6.1
Show that a star with a monochromatic flux of 3.63×10^{-20} erg sec^{-1} cm^{-2} Hz^{-1} has an AB magnitude of zero.

The Hubble Space Telescope Science Institute has adopted a similar zero-point system in which constant flux per unit *wavelength* interval has zero color index. In the ST system the monochromatic zero-point flux is 3.63×10^{-9} erg sec^{-1} cm^{-2} Å^{-1}. For the ST system

$$m_{ST}(\lambda) = -2.5 \log(F_\lambda) - 21.10. \qquad (6.2)$$

---
[1]An erg is CGS unit of energy equal to 10^{-7} J.

6.1.3 The Sloan Digital Sky Survey $ugriz$ System

The Sloan Digital Sky Survey (SDSS) is a project to produce images and spectra of a large fraction of the sky using a 2.5-m telescope at the Apache Point Observatory in New Mexico. The $ugriz$ system was developed for the CCD imagers used by the SDSS. The $ugriz$ system has five filters with peak response in the ultraviolet (u), green (g), red (r) and two $(i$ and $z)$ in the infrared. The system currently in use $(u'g'r'i'z')$ is actually slightly different than the original system $(ugriz)$ system. Given that the SDSS has measured photometry of nearly 200 million objects using the $u'g'r'i'z'$ system, $u'g'r'i'z'$ has become the most commonly used photometric system. Table 6.3 shows a comparison of $UBVRI$ system and the $u'g'r'i'z'$ system. Figure 6.2 shows the response functions for this system.

Table 6.3: Comparison of the $UBVRI$ and $u'g'r'i'z'$ system response function effective wavelengths (nm) and widths (nm). Data from Bessell[1] and Fukugita[11].

	$UBVRI$			$u'g'r'i'z'$	
	λ_{eff}	$\Delta\lambda$		λ_{eff}	$\Delta\lambda$
U	366.3	65	u'	356	46
B	436.1	89	g'	483	99
V	544.8	84	r'	626	96
R	640.7	158	i'	767	106
I	798.0	154	z'	910	125

6.2 PHOTOMETRIC DATA REDUCTION

In this section, we learn how to use a CCD image to determine a celestial object's flux in some standard photometric system. We will assume that you have already corrected the image using the techniques described in Section 5.3. We will concentrate on measuring the brightness of star-like sources, with a PSF that is approximately Gaussian in shape.

6.2.1 Aperture Photometry

Figure 6.3 is a negative image of a star with a Gaussian PSF. Our goal here is to estimate the signal from the star alone. One way to do this is

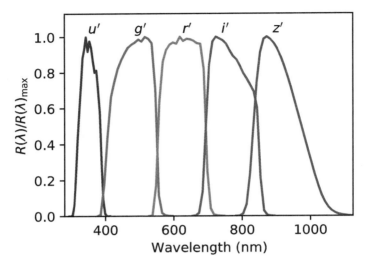

Figure 6.2: The u', g', r', i', and z' response functions for a typical CCD. The curves represent the expected total quantum efficiencies of the camera plus telescope. (Data from Smith et al.[20].)

to simply add all of the signal in a circular region or aperture around the star's center and then subtract the contribution of the sky background. This procedure is called **aperture photometry**. The details of the reduction depends on the detector used, but the general procedure for estimating the signal from a star consists of three steps: finding the center of the star's PSF, estimating the sky background signal from an annular region around the star's center, and then computing the signal contributed by the star in a circular aperture around the star's center.

Finding the PSF center

One of the simplest ways to determine the center of the PSF is to compute the centroid of a square region (sometimes called the centering box) around the brightest pixel in the star's image as described in Section 5.4.1. For well-sampled PSFs with good SNR, this technique gives results good to within a small fraction of a pixel. There are a number of more sophisticated techniques that give better results, but a fraction of a pixel is usually good enough.

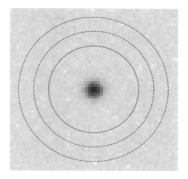

Figure 6.3: Synthetic image of a star with a 5 pixel FWHM Gaussian PSF. The smaller solid circle indicates the photometry aperture. The aperture has a radius of 3 times the PSF or 15 pixels. The dotted circles mark the inner and outer radii of the annular region used to determine the sky background.

Sky background estimation

The signal around the center of the star includes sky background as well as stellar flux. We need to remove the sky background signal to get an accurate estimate of the star's flux. A straightforward way to estimate the sky background signal is to sum the signal from an annular region centered on the star and divide by the number of pixels in the annulus. The dashed lines in Figure 6.3 show the inner and outer radii of the annulus. You must choose the inner radius of an annulus to be far enough from the center of the PSF that the contribution from the star's signal is insignificant. Generally picking the radius of the inner annulus, $r_{\text{inner}} \gtrsim 3 \cdot \text{FWHM}$ is a reasonable choice. The background signal will finally be subtracted from the source signal, so we need to make sure the background's uncertainty is small. To make sure the background uncertainty is small, you need to pick a radius for the outer annulus to include enough pixels for a good statistical determination. One often-used rule of thumb is to choose the outer annulus so that the total number of pixels in the annulus is about three times the number of pixels in the aperture.

The procedure described above computes the sky background per pixel from the mean signal in the annulus. If there is another object in the annulus, the mean will be higher than the actual background value. The median value is less affected by the outlying high pixel values. Many sophisticated astronomical image process programs also allow you to select the option of applying **three-sigma clipping.** The three-sigma clipping algorithm computes the mean (or median) background signal per pixel $\bar{B}_{\mathrm{sky}}$, and the standard deviation, σ, of the pixel values in the annulus. It then removes all pixels outside the range from $\bar{B}_{\mathrm{sky}} - 3\sigma$ to $\bar{B}_{\mathrm{sky}} + 3\sigma$ from the sample, and finally recomputes the mean (or median). This technique not only eliminates most of the contamination from an object in the annulus, but also removes cosmic ray hits, and bad pixels.

Star signal estimation

The simplest technique for estimating the star's signal is to simply add up pixel values of all the pixels in a circular aperture centered on the star and then subtract the background signal in the aperture. Let's define $S_\star$ to be the sum of the pixel values in the aperture,

$$S_\star = \left(\sum_x \sum_y S_{xy} \right), \tag{6.3}$$

where the sums extend only over pixels inside the aperture. The sky-subtracted signal from the star alone is then

$$I_\star = S_\star - N_{\mathrm{pix}} \bar{B}_{\mathrm{sky}}, \tag{6.4}$$

where N_{pix} is the number of pixels in the aperture, and $\bar{B}_{\mathrm{sky}}$ is the estimated background per pixel. The aperture radius r_{ap} needs to be large enough that the aperture includes most of the light from the source. A radius $r_{\mathrm{ap}} = 3 \cdot \mathrm{FWHM}$ includes essentially 100% of a Gaussian PSF and is a reasonable choice.

If you write a program to sum the pixel values in the aperture, you will have to decide how to deal with pixels on the edge of the aperture. One approach would be to require all the pixels in the sum to be fully inside the aperture; the program would just ignore any pixels what are intersected by the aperture radius. This works fine for stars whose FWHM in pixels is large, but for small PSFs compared to the pixel size one pixel might contain a significant amount of flux. A better choice might be to multiply the signal in an intersected pixel by the fraction of the area of the pixel included in the aperture. This is the technique used by most sophisticated aperture photometry programs.

The procedure described above is the simplest way to perform photometry of unresolved or compact objects. Most astronomical image processing packages have the option of using much more sophisticated algorithms. You will need to use these if you are doing photometry of extended objects or in crowded fields with many stars. A technique called PSF fitting fits a synthetic or sampled PSF to point sources and extracts the background and source signal from the fit. PSF fitting works much better than aperture photometry when the background isn't flat, the field is crowded, or you need to determine the brightness of an extended object. The companion website[2] lists several astronomical image processing packages used for photometry by professional astronomers.

6.2.2 Count Rate and Instrumental Magnitudes

The final step is to determine the photon count rate from the star. The value of $I_\star$ computed using equation (6.4) is the number of counts in ADU collected by the CCD. The number of photons $n_\star = gI_\star$, where g is the gain of the CCD in photons/ADU. The count rate is

$$\dot{n}_\star = \frac{n_\star}{t_{\text{exp}}} = \frac{gI_\star}{t_{\text{exp}}}, \tag{6.5}$$

where t_{exp} is the exposure time. We compute the **instrumental magnitude** using equation (3.24),

$$m = -2.5 \log(\dot{n}_\star) + C. \tag{6.6}$$

The instrumental magnitude is just a convenient way expressing the measured count rate. The value of C is arbitrary, but is typically chosen so that the instrumental magnitude is approximately equal to the object's magnitude in a standard photometric system. The value of C is typically in the range from 24 to 25 for a one-meter telescope. It is somewhat larger for large telescopes and smaller for smaller aperture telescopes.

Problem 6.2
Use the information in Table 3.2 to estimate a value for C that will make the measured instrumental magnitudes of stars close to their B magnitudes when observed using a 1-m telescope. Assume that the

---

[2]https://mshaneburns.github.io/ObsAstro/

transmissivity of the optics is $T_o = 0.8$, and the filter transmissivity is $\phi' = 0.5$. Most astronomical detectors have poor blue efficiency so assume $e' = 0.3$. Ignore the effect of atmospheric extinction.

Problem 6.3

Use equations (6.5) and (6.6) to show that you can also write the instrumental magnitude as

$$m = -2.5\log(I_\star) + C'. \tag{6.7}$$

Find the new constant C' in terms of C, g, and t_{exp}.

6.2.3 Photometric Uncertainty

Every measurement has some uncertainty. We can determine the uncertainty in the instrumental magnitude by tracing back through each step of the processing and applying the error propagation techniques described in Appendix B.7.

We start by applying equation (B.27) for error propagation to equation (6.6),

$$\delta m = \frac{2.5}{\log(10)} \frac{\delta \dot{n}_\star}{\dot{n}_\star} \approx \frac{\delta \dot{n}_\star}{\dot{n}_\star}. \tag{6.8}$$

If we assume there is no uncertainty in the exposure time then

$$\delta m = \frac{\delta n_\star}{n_\star} \tag{6.9}$$

Given that $n_\star = gI_\star$, equation (6.4) implies that

$$n_\star = gS_\star - N_{\text{pix}}\, g\bar{B}_{\text{sky}}. \tag{6.10}$$

The error propagation equation (B.27) gives

$$\delta n_\star^2 = (g\,\delta S_\star)^2 + N_{\text{pix}}^2 (g\,\delta \bar{B}_{\text{sky}})^2.$$

If we have chosen a sufficiently large annulus to compute the sky background, then $(\delta S_\star)^2 \gg (N_{\text{pix}}\, \delta \bar{B}_{\text{sky}})^2$, so $\delta n_\star = g\,\delta S_\star$. This calculation illustrates the importance of choosing a large enough sky annulus and gives a quantitative way of deciding how big it should be.

Now we need to compute $g\delta S_\star$. The value of $S_\star$ is computed by summing the pixel values of all pixels in the aperture (equation (6.3)). This sum includes both the number counts from the source as well as the sky background. The uncertainty in the number of counts is subject to two sources of uncertainty, photon counting noise and detector noise (see Section 3.3.2). The photon counting noise δn_γ is governed by Poisson statistics so $\delta n_\gamma = \sqrt{n_\gamma} = \sqrt{g\,S_\star}$ where n_γ is the number of photons collected from the source and the sky background. (If the detector has significant dark current, n_γ would also include the dark current. I've assumed the dark current is negligible for this calculation.) If the detector is a CCD then the detector noise contribution $\sigma_{\text{det}}^2 = N_{\text{pix}}\sigma_{\text{CCD}}^2$, where σ_{CCD} is the CCD read noise.

Adding the photon counting noise in quadrature with the detector noise gives

$$\delta n_\star = \sqrt{g\,S_\star + N_{\text{pix}}\sigma_{\text{CCD}}^2}. \tag{6.11}$$

By using equations (6.9), (6.10), and (6.11) you can show that the uncertainty in the instrumental magnitude is

$$\delta m = \frac{\sqrt{g\,S_\star + N_{\text{pix}}\sigma_{\text{CCD}}^2}}{gS_\star - N_{\text{pix}}\,g\bar{B}_{\text{sky}}} \tag{6.12}$$

The pixel values S_{xy} in the sum used to compute $S_\star$ [equation (6.3)] are the values after the raw CCD data has been had the bias and dark current subtracted, and the efficiency map applied, from equation (5.6) this implies that

$$S_{x,y} = \frac{R_{x,y} - D_{x,y}\,t_{\text{exp}} - Z_{x,y}}{E_{x,y}}, \tag{6.13}$$

where $R_{x,y}$ is the raw pixel value, $D_{x,y}$ is the dark current, $Z_{x,y}$ is the bias frame pixel value, and $E_{x,y}$ efficiency frame value. For the calculation above, we assumed that the uncertainties in the calibration images are small compared to uncertainties in data pixel values $R_{x,y}$. Hence, the uncertainty in $R_{x,y}$ dominates all the other errors and $\delta S_{x,y} = \delta R_{x,y}/E_{x,y}$. By design, $E_{x,y} \approx 1$ so $\delta S_{x,y} = \delta R_{x,y}$. It is essential to make sure the error in the calibration images is small to get a good estimate of the uncertainty using equation (6.12).

6.2.4 PSF Fitting

Aperture photometry is simple and works well if the stars are well isolated from one another. The technique gives poor results if the field is crowded because the stars PSFs overlap. Some of the most interesting astrophysics requires photometry in crowded fields. These include globular clusters, distant open clusters, and stars in external galaxies. Even some isolated objects like supernovae in distant galaxies or stars in star-forming regions have backgrounds that aren't handled very well using aperture photometry techniques. PSF fitting avoids problems inherent in aperture photometry. The technique typically starts by automatically finding potential stars in the images. A a two-dimensional analytic or empirical model for the PSF with background is then fit to the image. The fit parameters include the total signal $S\star$ and the background which are used to compute the instrumental magnitudes and uncertainties of all the stars. For very crowded fields with stars with a large range of brightnesses, the process can be used iteratively. After the first set of stars in found and fit, the fit is subtracted from the image and the process is repeated to get photometry for the fainter stars. PSF fitting is a complex process, but essentially all sophisticated image processing packages implement some form of PSF fitting photometry.

6.3 DIFFERENTIAL PHOTOMETRY

Suppose you are observing a star and wish to determine if it is a variable. In this case, you don't need to tie the star's brightness to a standard photometric system. We only need to compare this star's brightness with some reference star that we know isn't variable. This sort of comparison is called **differential photometry**. If both of the stars are in the same image, this is easy to do. Suppose you use one of the photometry techniques described above to determine the instrumental magnitude of the target star to be m_T and a comparison star to be m_C. By plotting the difference in magnitudes

$$m_T - m_C = -2.5 \log \left(\frac{I_T}{I_C} \right) \qquad (6.14)$$

over a period of time, we can determine if the target is variable. Sometimes we might not know if the comparison star is variable, in this case we can use two or more stars for the comparison and plot the difference in magnitudes between the two comparison stars as well. Figure 6.4 shows an example. The top panel of the plot shows that $m_T - m_{C1}$ varies periodically with an amplitude of about 0.2 mag. The bottom panel shows that $m_{C1} - m_{C2}$ has a random scatter of about 0.03 mag. If the scatter in $m_{C1} - m_{C2}$ had been comparable to the amplitude of

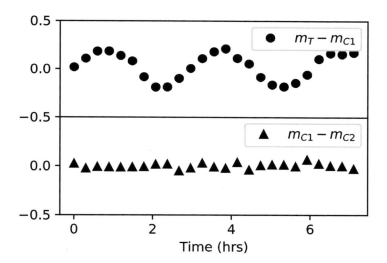

Figure 6.4: A plot of the light curve for a variable star. The top panel shows difference between instrumental magnitudes of the variable star and a comparison star. The bottom panel shows the difference in instrumental magnitudes of not non-variable comparison stars. Note that the vertical scales for both panels is the same. The plots have also been adjusted so that the average of the differences in magnitudes are zero.

$m_T - m_{C1}$ or showed an increasing or decreasing trend we would know that at least one of the comparison stars was suspect.

If there is a standard star in the same image as the target, we can sometimes use differential photometry to tie the magnitude of the target star to a standard photometric system,

$$m_T = m_S - 2.5 \log \left(\frac{I_T}{I_S} \right), \tag{6.15}$$

where m_S is the know magnitude of the standard star. This technique works well if the spectra of the standard and target are similar. If not, then we must transform m_T to the standard system using the techniques discussed in the next sections.

6.4 ABSOLUTE PHOTOMETRY

The goal of absolute photometry is to determine the flux density or magnitude of an astronomical object in some standardized bandpass. Absolute photometry is needed if for example you wish to construct a color-magnitude diagram or use the inverse square law to measure distances. Determining the magnitude in some standard system is complicated by at least three factors (i) spectral differences between the target and standard stars, (ii) departures of the observing system from the standard systems, and (iii) atmospheric extinction. Atmospheric extinction has potentially the largest effect.

6.4.1 Atmospheric Extinction

As light passes through the Earth's atmosphere some of the photons are absorbed or scattered. In either case these photons don't make it to the detector. The amount of scattering or absorption is wavelength dependent. Essentially all light with wavelengths less than about 300 nm is scattered or absorbed. Although there are a few absorption lines toward the red end of the visible spectrum, the opacity of the atmosphere decreases smoothly from 300 nm to about 900 nm. Molecular absorption lines dominate the spectrum from about 900 nm to about 5 μm. There is another low-opacity window in the IR from about 5 μm to 8 μm.

The amount of extinction depends on the amount of atmosphere between the source and the detector. The light passes through more atmosphere if we observe the source when it is closer to the horizon. Let's use the simple slab atmosphere model illustrated in Figure 6.5 to

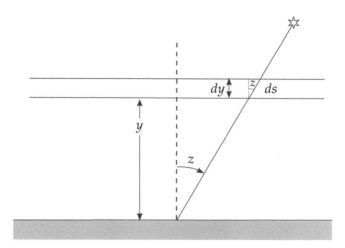

Figure 6.5: A slab atmosphere model. The star is seen at zenith angle z. The total hight of the slab is h. The absorption coefficient $\alpha(\lambda, y)$ depends only on wavelength and the height y above the ground.

quantify the extinction. The star is observed at the zenith angle z. As light from the star passes down through the atmosphere some of the photons are scattered or absorbed. In traversing the path ds we will assume the fraction of flux absorbed is

$$\frac{dF_\lambda}{F_\lambda} = -\alpha(\lambda, y)ds, \qquad (6.16)$$

where $\alpha(\lambda, y)$ is the fraction of light absorbed per unit distance. The value of $\alpha(\lambda, y)$ depends on wavelength and the density of the atmosphere. In this simple model we assume the atmosphere consists of slabs of constant density so that $\alpha(\lambda, y)$ depends only on height above the ground y. Using the fact that $ds = \sec(z)dy$ you can integrate equation (6.16) to find the observed flux $F_{\lambda,\text{obs}}$ in terms of the flux above the atmosphere $F_{\lambda,0}$,

$$F_{\lambda,\text{obs}} = F_{\lambda,0} \exp\left(-\sec(z)\int_0^h \alpha(\lambda, y)dy\right), \qquad (6.17)$$

where h is the hight of our model atmosphere. The integral $\int_0^h \alpha(\lambda, y)dy = \tau_\lambda$ is the optical depth of the atmosphere. (See Chapter 9 of Carrol and Ostlie [6] for a complete discussion of optical depth.) The **airmass** X is defined to be the effective number of atmospheres one is looking through. For our simple slab atmosphere model $X = \sec(z)$. When looking straight up toward the zenith $z = 0$ and $X = 1$; we are looking through one atmosphere. When our object is at $z = 60°$, $X = 2$ and the object's light would suffer two atmospheres of extinction. The expression $X = \sec(z)$ is an approximation that doesn't include the curvature of the Earth or refraction by the atmosphere. Various more exact empirical models have been developed by fitting extinction measurements at different zenith angles. One commonly used model developed by Hardie [12] uses

$$
\begin{aligned}
X =\ & \sec z - 0.0018167(\sec(z) - 1) - 0.002875(\sec(z) - 1)^2 - \\
& 0.0008083(\sec(z) - 1)^3. \qquad (6.18)
\end{aligned}
$$

This model is introduces a very small correction that is only important for $z > 60°$, but gives a more accurate value for X up to about $z = 85°$. Regardless of which equation we use for airmass, we can rewrite equation (6.17) as

$$F_{\lambda,\text{obs}} = F_{\lambda,0}\, e^{-\tau_\lambda X}. \qquad (6.19)$$

In Chapter 3, we showed that the flux F measured in a particular bandpass was

$$F = \int_0^\infty R(\lambda) F_\lambda d\lambda$$

[equation (3.4)]. If we now take atmospheric extinction into account using equation (6.19) we get

$$F_{\text{obs}} = \int_0^\infty R(\lambda) F_{\lambda,0} e^{-\tau_\lambda X} d\lambda.$$

If the extinction factor is a relatively smooth function over the bandpass, then we can pull that factor out of the equation to get

$$F_{\text{obs}} = e^{-\tau_{\text{eff}} X} \int_0^\infty R(\lambda) F_{\lambda,0} \, d\lambda = e^{-\tau_{\text{eff}} X} F_0, \tag{6.20}$$

where τ_{eff} is an effective optical depth over the entire bandpass. Here, F_0 is the flux that would be measured from the object above the atmosphere. We can rewrite this equation in magnitudes,

$$m = m_0 + k \, X, \tag{6.21}$$

where k called the **extinction coefficient**, m is the observed magnitude of the star, m_0 is the magnitude of the star that would be observed from above the atmosphere.

Problem 6.4
Show that $k = 2.5 \tau_{\text{eff}} \log(e)$.

Extinction coefficients depend on atmospheric conditions and can change with time. Nights during which the transparency doesn't change significantly and is uniform across the sky are said to be photometric. During photometric nights you can determine extinction coefficients for different bandpasses by observing a standard star for a few hours before and after it transits the meridian. Figure 6.6 shows R-band measurements taken for a star along with a fit of equation (6.21). The slope of the fit line gives the extinction coefficient k and the intercept m_0 is the instrumental magnitude of the star above the atmosphere. Once you have determined the extinction coefficient, you compute the magnitude of any unknown star above the atmosphere from the instrumental magnitude and the airmass at the time of observation.

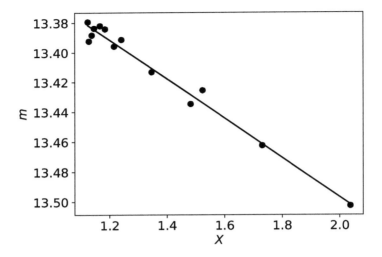

Figure 6.6: R-band instrumental magnitudes plotted versus airmass for a standard star. The solid line is a linear fit to the data.

Problem 6.5
The R-band extinction determined from the data in Figure 6.6 gives an extinction coefficient $k_R = 0.141 \pm 0.008$. Suppose you observe another star at a zenith angle of 18.5° to have an instrumental magnitude $m = 16.01 \pm 0.02$. What is magnitude of this star above the atmosphere?

Atmospheric extinction depends strongly on wavelength in the visible band ranging from about 0.5 magnitudes in the U-band to often less than a tenth of that in the I-band. Figure 6.7 shows one set of measurements of the extinction coefficients in the Johnson-Cousins B, V, R, and I bands.

In almost all cases in which we need to do absolute photometry, we need to make measurements in several bandpasses. For example, suppose we need U, B, and V photometry. We could observe a standard star at a full range of zenith angles and compute three first-order extinction coefficients k_v, k_b, and k_u by plotting the standard star's instrumental magnitudes v, b, and u respectively vs. the airmass X. A linear fit would

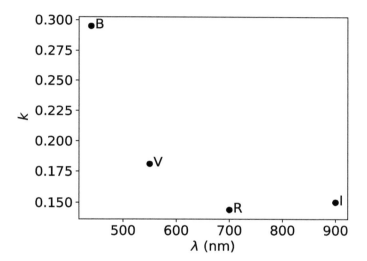

Figure 6.7: Extinction values for the Johnson-Cousins B, V, R, and I filters. The observations were done using a 0.41-m telescope at an observatory in the middle of Colorado Springs on a marginally photometric night. The larger extinction toward the blue part of the spectrum is apparent.

give the extinction coefficients (the slopes of the fit lines). The extinction corrected instrumental magnitudes for the stars of interest would then be then calculated from the equations,

$$u_0 = u - k_u X, \tag{6.22}$$
$$b_0 = b - k_b X, \tag{6.23}$$
$$v_0 = v - k_v X, \tag{6.24}$$

where the 0 subscript indicates the above-atmosphere magnitudes and X is the airmass. However, extinction over a single filter bandwidth is actually color dependent so for a very precise work we could compute a second-order fit by letting $k_u = k'_u + k''_u(b_0 - v_0)$, $k_b = k'_b + k''_b(b_0 - v_0)$, and $k_v = k'_v + k''_v(b_0 - v_0)$. Determining second order coefficients requires a significant amount of observing time. Fortunately, second order coefficients are typically small and can be ignored.

6.4.2 Standard System Transformation Coefficients

The magnitudes of stars in any one of the standard systems depends on the spectrum of the star and perhaps even differences between the ideal standard system and the system used for the observations. Again, the transformations are often small and fortunately they don't change from night to night. Many large observatories have devoted a substantial amount of observing time to determining the coefficients and even provide software tools for making the correction. Below I describe a very simple example of using first-order transformation for a set of U, B, and V band observations.

The first step is to determine the extinction-corrected instrumental magnitudes of the stars. Without including transformation you can determine the magnitudes and color indices your target stars using

$$V = v_0 + (V_S - v_{0S}), \tag{6.25}$$
$$B - V = (b_0 - v_0) + [(B - V)_S - (b_{0S} - v_{0S})], \tag{6.26}$$
$$U - B = (u_0 - b_0) + [(U - B)_S - (u_{0S} - b_{0S})], \tag{6.27}$$

where the S subscript indicates the values for the standard star. Notice that we are computing the color indices $B - V$ and $U - B$ rather than U, B, and V independently. This is typically the way photometric data for stars is presented. The values $u, b, v, u_S, b_S,$ and v_S are all extinction corrected instrumental magnitudes that are determined from the observations. V_S, $(B - V)_S$, and $(U - B)_S$ are the accepted values for the magnitudes and color indices of the standard star. See Landolt [14] for lists of $UBVRI$ standard stars and Smith et al.[20] for a list of the standard stars that define the $u'g'r'i'z'$ system.

The first order transformation the equations for transforming from the extinction corrected instrumental magnitudes to the standard UBV system are

$$
\begin{aligned}
(B - V) &= \mu(b_0 - v_0) + C_{bv}, & (6.28) \\
V &= v_0 + \epsilon(B - V) + C_v, & (6.29) \\
(U - B) &= \psi(u_0 - b_0) + C_{ub}. & (6.30)
\end{aligned}
$$

Observations of many standard stars are used to determine the transformation coefficients ϵ, μ, ψ, and the Cs. For example, to determine ϵ you would plot $V - v_0$ vs. $B - V$ for many standard stars of different colors. The slope of the line would give ϵ and the intercept would give C_v.

SUPPLEMENTARY PROBLEMS

Problem 6.6 The table below summarizes several measurements of the magnitude of a star at different air masses. Fit equation (6.21) to these data and estimate the extinction coefficient and the star's magnitude above the atmosphere. Don't forget to include uncertainty estimates in your answers. You can download a CSV file with these data from the companion website.[3]

Airmass	Magnitude
1.000	14.33
1.108	14.41
1.227	14.45
1.359	14.40
1.506	14.42
1.668	14.52
1.848	14.45
2.047	14.57
2.268	14.56
2.512	14.57

Problem 6.7 Fit the data in the table of Problem 6.6, but this time

---

[3]https://mshaneburns.github.io/ObsAstro/

compute the reduced χ^2 for the fit assuming the uncertainty in the magnitude measurements is 0.03. Do you think 0.03 is a good estimate of the uncertainty in the magnitude? Explain your reasoning.

Spherical Trigonometry

The shortest distance between any two points on the surface of a sphere can be connected by a great circle. A **great circle** is the intersection of the surface of a sphere and a plane that also passes through the center of the sphere. The intersection of three great circles on the surface of a sphere forms a **spherical triangle** (see Figure A.1). The **law of cosines**

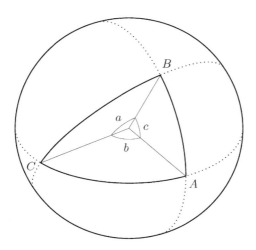

Figure A.1: A spherical triangle is formed by three great circles. The angles a, b, c, A, B, and C are related by the law of sines and the law of cosines. (Source: Wikimedia Commons)

for spherical triangles can be written in two ways,

$$\cos a = \cos b \cos c + \sin b \sin c \cos A \qquad (A.1)$$

DOI: 10.1201/9781003203919-A

or
$$\cos A = \cos B \cos C + \sin B \sin C \cos a. \qquad \text{(A.2)}$$

The **law of sines** for spherical triangles is

$$\frac{\sin a}{\sin A} = \frac{\sin b}{\sin B} = \frac{\sin c}{\sin C}. \qquad \text{(A.3)}$$

For derivations of these equations and a complete description of spherical trigonometry see Smart [19, Chapter 1].

Data Analysis

B.1 WHAT IS UNCERTAINTY?

Experimental measurements never yield exact results. For example, suppose you were asked you to measure the length of a small table with a meter stick. You would carefully align one end of the meter stick with one edge of the table then look at the other edge to read off the length. You read off the length by looking at the marks on the meter stick and determining which one lines up with the edge of the table. Is there uncertainty in the measurement? Yes, because you can't read the scale more finely than about a millimeter. The smallest marks on a meter stick are usually one mm apart. The best you could probably do would be to say that the length of the table is in some range.

Suppose you make the measurement described above and find the length to be between 61.2 and 61.4 cm. Your instructor would like you to write a short lab report on your measurement. (I know, instructors ask you to do some weird things.) How would you report the length? You could say "the length of the table is between 61.2 and 61.4 cm." This would be correct, but scientists have a convention for specifying the results of a measurement; we specify a best estimate plus or minus an uncertainty. In this case, the best estimate of the length would be the value in the middle of the range or 61.3 cm. The uncertainty specifies how much bigger or smaller the length could be or in this case 0.1 cm. In this example, the range 61.2 to 61.4 cm should be specified as 61.3 ± 0.1cm.[1] In general,the result of a measurement of some quantity x is stated as

$$(\text{measured value of } x) = x_{\text{best}} \pm \delta x, \tag{B.1}$$

---

[1]The $\pm$ symbol is read as "plus or minus".

where x_{best} is the best estimate of x and δx is the uncertainty in the estimate. We'll learn some sophisticated ways of computing x_{best} and δx later, but if our measurement of x is between x_{max} and x_{min}, then

$$x_{best} \approx \frac{x_{max} + x_{min}}{2}, \tag{B.2}$$

and

$$\delta x \approx \frac{x_{max} - x_{min}}{2} \tag{B.3}$$

give crude estimates.

Knowing the uncertainty is essential to testing scientific theories. For example, suppose a new theory of stellar evolution predicts the mass of a star in a binary system is $2.1\,M_\odot$. Suppose you measure the mass and find it to be $1.7\,M_\odot$. Do these numbers agree? It depends on the uncertainty. If the uncertainty in our measurement was $\pm 0.5\,M_\odot$ then our measurement is consistent with the theory, but if the uncertainty was $\pm 0.1\,M_\odot$ we would have to conclude that the theory is inconsistent with the measurement. A measurement without an estimated uncertainty usually isn't very useful. Astronomers often spend as much or more time determining the uncertainty in a measurement as they do determining the result of the measurement.

B.2 REPORTING UNCERTAINTIES

Let's explore some conventions scientists use to report their result. Uncertainty estimates are, after all, just estimates so they should not be stated with too much precision. In our table example, it wouldn't make any sense to quote our measured length as 61.3 ± 0.1239856 cm. It simply isn't possible to know the uncertainty to seven significant figures. Uncertainties are usually quoted to one or two significant figures. In this book we will use the following convention:

> If the most significant digit in the uncertainty is a '1' we will round the uncertainty to two significant figures, otherwise we will round uncertainties to one significant figure.

According to this convention the uncertainty 0.1239856 would be rounded to 0.12. If the computed uncertainty had turned out to be 0.85438, it would be rounded to 0.9.

Once the uncertainty has been rounded, we must also consider the number of significant figures to keep in the best estimate. A statement like "the measured speed = 6056.32 ± 3 m/s" is obviously ridiculous. The uncertainty of 3 m/s means that the digit '6' in the fourth place of 6056.32 might really be as small as '3' or as large as '9'. Clearly the trailing digits '.32' have no significance at all, and should be rounded off. The proper way to state the result is 6056 ± 3 m/s.

The procedure for determining how to report a measured value with its uncertainty is to first use the convention above to determine the number of significant figures to keep in the uncertainty and then round the best estimate value to match the number of significant figures in the uncertainty. The least significant figure in any measured value should be of the same order of magnitude (in the same decimal position) as the uncertainty.

For example, suppose you have three measured values of a time interval: 2.335, 2.437, and 2.270 seconds. Using your calculator to compute the uncertainty according to equation (B.3) you would get 0.0835. Using the convention for rounding uncertainties gives 0.08. Using your calculator again to compute the average of the three values would give 2.347333. Using the convention for rounding measured values to match the uncertainty gives 2.35. You would report your best estimate of the time interval as 2.35 ± 0.08 seconds.

There are two other things to keep in mind when reporting measured values. The first is that since the uncertainty and the best estimate both have the same units it is clearer to write the result as 2.35 ± 0.08 seconds rather than 2.35 seconds ± 0.08 seconds. Second, we will often measure numbers that are reported in scientific notation. Suppose we measured a distance to be 1.61×10^5 AU with an uncertainty of 5×10^3 AU. The clearest way to report this is $(1.61 \pm .05) \times 10^5$ AU rather than $1.61 \times 10^5 \pm 5 \times 10^3$ AU.

Problem B.1
The table below is a collection of measurements and estimated uncertainties. Write them in the proper format.

Measurement	Uncertainty
35.9824 m/s	0.113 m/s
1.022089×10^{19} kg	6.321×10^{15} kg
0.0012 A	0.1935 A
1235 AU	211 km

B.3 ESTIMATING UNCERTAINTIES

There are a huge number of different ways to make a measurement which means there are just as many different ways to estimate the uncertainty in the measurement. In fact you can often estimate the uncertainty in a number of different ways and it is common when making a complex measurement to use several different methods to estimate the uncertainty in order to cross check the estimates.

In the example of measuring the length of a table, we got our uncertainty estimate by simply estimating the accuracy to which we thought we could read the meter stick. This is a quick and simple way to get an estimate. Estimates of this kind are somewhat subjective, but sometimes it is the best that you can do. A better, but more time consuming technique, would be to have several different experimenters make the measurement then combine their results somehow. We'll discuss exactly how we might combine their results in Section B.5.

Most modern lab equipment is digital. A modern voltmeter is a good example. Suppose we used a voltmeter to measure the voltage across a battery and the digital display read 1.23 volts. What is the uncertainty in this measurement. If we assume the voltage of the battery doesn't change during our measurement and that the voltmeter is calibrated properly, the accuracy of the measurement is given by the accuracy of the display. The display would read the same if the actual voltage where anywhere in the range from 1.225 to 1.234 volts. Using the equation (B.2) for the best estimate, we get 1.230 volts and using equation (B.3) for the uncertainty gives 0.005 volts. Our best estimate of the battery voltage is then 1.230 ± 0.005 volts. In other words, when reading a digital display the uncertainty is $1/2$ the least significant digit that can be displayed.

B.4 SYSTEMATIC VERSUS RANDOM ERRORS

In our voltmeter example we assumed that the voltmeter was properly calibrated. If it wasn't, then our best estimate wouldn't be correct. A calibration error like this is an example of a **systematic error**. Systematic errors are errors associated with a flaw in the equipment or in the design of the experiment. Systematic errors cannot be estimated by repeating the experiment with the same equipment. In the battery example, the best way to deal with the systematic error would be to recalibrate the voltmeter. If this isn't possible, then we would have to somehow make an estimate of the possible size of the systematic error and include it in our uncertainty estimate. Systematic errors are insidious because the experimenter usually doesn't know they are present. If they did they would correct the flaw before doing the experiment.

Hubble's original estimate of his eponymously named constant was $500 \, \mathrm{km \, s^{-1} Mpc^{-1}}$. This is about seven times larger than the currently accepted value of about $70 \, \mathrm{km \, s^{-1} Mpc^{-1}}$. This huge discrepancy was due to a systematic error on Hubble's part. He determined the distances to galaxies by identifying Cepheid variable stars, but he misidentified the population of Cepheids and hence got the distances wrong. He identified the linear relationship between distance and velocity, but his actual number for the proportionality was incorrect. This is a classic example of an unidentified systematic error.

Many measurements involve uncertainties that can't be estimated by reading a scale or a digital display. For example, if I measure a time interval using a digital stopwatch, the main source of uncertainty isn't usually from the accuracy of stopwatch display, but is from the variability of my reaction time. A good way to estimate the uncertainty in this case would be to make repeated measurements. Errors that can be reliably estimated by repeating measurements are called **random errors**.

B.5 STATISTICAL ANALYSIS OF RANDOM ERRORS

Table B.1 gives the 24 distance measurements for a nearby galaxy. How

Table B.1: Distance measurements to a nearby galaxy in kpc.

741	804	780	714	810	760	778	741	737	752	780	814
742	788	760	811	848	768	742	741	786	764	696	783

should we determine the best estimate of the distance to the galaxy along with the uncertainty? Let's first deal with determining the best estimate of the distance first. One way to estimate the distance would be to take the mean or average of all 24 measurements. The **mean** of a set of N values $\{y_1, y_2, ..., y_N\}$ is denoted $\bar{y}$ and is defined as

$$\bar{y} = \frac{1}{N} \sum_{i=1}^{N} y_i. \tag{B.4}$$

The mean for the data set in Table B.1 is 768.3 kpc. Figure B.1 shows a histogram of the data. The mean is near the center of the distribution of

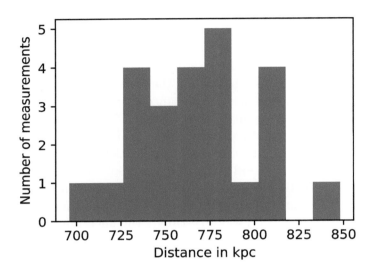

Figure B.1: Histogram of the distance measurements from Table B.1.

data points. In this case the mean is a good best estimate of the distance to the galaxy, but what if the distribution of distance measurements looked like the histogram shown in Figure B.2? In this case a single, probably spurious, outlying distance measurement at 400 kpc draws the mean away from the central value of the distribution. In this case the median is probably a better estimate of the distance to the galaxy. The **median**, $\tilde{y}$ is determined by sorting the data from lowest to highest and picking the middle value when the sample size is odd, or averaging the two numbers closest to the middle when the sample size is even. For example suppose five measurements of y yield the numbers $\{1, 8, 4, 3, 5\}$, then sorting the numbers gives $\{1, 3, 4, 5, 8\}$ and the median is $\tilde{y} = 4$.

The median is a good statistic when outliers are suspected, but it is difficult to compute because it requires sorting the data. A simpler statistic is the **mode** which is just the most frequent value in the data. The mode for the histogram shown in Figure B.2 is essentially identical to the median, both of these statistics are less affected by the outlier than the mean.

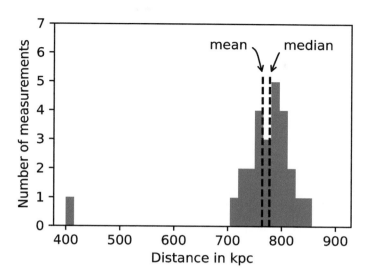

Figure B.2: Histogram of distance measurement with one outlier at 400 kpc. The mean is biased toward the outlier more than the median. The mode is the same as the median in this histogram.

Problem B.2

An astronomer measures the duration of the twenty successive eclipses of a binary star system and gets the following results:

15.5,	15.2,	15.6,	15.7,	14.9,	14.7,	14.6,	15.4,	15.4,
14.8,	14.9,	15.1,	14.2,	14.5,	14.8,	14.9,	15.6,	14.1,
15.7,	14.7,							

where all times are measured in days. Compute the mean, median and mode of these data.

The uncertainty in the measurement is related to the width of the distribution of data points. The crude estimate of equation (B.3) is very sensitive to outlying points. A better statistic that more accurately estimates the width of the distribution, and as will learn in the next section has an important interpretation in terms of probabilities, is the **standard deviation**. The **population standard deviation** is the root-mean-square deviation of the data values from the mean,

$$s_p = \sqrt{\frac{1}{N} \sum_{i=1}^{N} (y_i - \bar{y})^2}, \tag{B.5}$$

and is relevant when the *entire* set of all possible data values is known. This is seldom the case in science. We are usually trying to *estimate* the width of the distribution from a sample of the population. In this case the relevant statistic is the **sample standard deviation**

$$s = \sqrt{\frac{1}{N-1} \sum_{i=1}^{N} (y_i - \bar{y})^2}. \tag{B.6}$$

The population standard deviation differs negligibly from the sample standard deviation when we have more than a few data points. See Bevington and Robinson [3] for a detailed explanation for the difference between the sample and population standard deviations.

The standard deviation is a quantitative measure of the width of the distribution of data values, it shouldn't change significantly as we take more data if there is the same amount of uncertainty in each measurement. However, our intuition tells us that mean value derived from the data should be more accurate as we take more data. This is in fact the case. The standard deviation is a quantitative measure of the uncertainty in a *single* measurement. The uncertainty in the mean is estimated by computing the **standard deviation of the mean**, σ_{mean}, which is computed by dividing the sample standard deviation by $\sqrt{N}$,

$$\sigma_{\text{mean}} = \frac{s}{\sqrt{N}}. \tag{B.7}$$

For the data of Table B.1, $\bar{y} = 768.3$ kpc, the median is essentially the same. There aren't any outliers so we can take the mean as the best estimate of the distance, $y_{\text{best}} = \bar{y} = 768.3$ kpc. The standard deviation of the data is $s = 34$ kpc. The estimate for the distance to the galaxy is determined from the mean of the data so the uncertainty in the distance is the standard deviation of the mean, $\delta y = \sigma_{\text{mean}} = s/\sqrt{24} = 6.9$ kpc. Rounding the uncertainty to one significant figure and matching the significant figures in the mean gives the best estimate of the distance as 768 ± 7 kpc.

Problem B.3

(a) Compute the sample standard deviation and the standard deviation of the mean for the data in Problem B.2.

(b) State the best estimate for the eclipse duration with its associated uncertainty.

B.6 PROBABILITY DISTRIBUTIONS

If we continued to make additional measurements of the distance to the galaxy, each measurement would give a slightly different value for the distance. The distance measurement is subject to random variations. Variables that are subject to random variations due to chance are called **random variables**. Of course there is a higher probability of getting some values of distance and a lower probability of getting others. A **probability distribution** is a function that describes how likely we are to get a particular value of a random variable. If the random variable, call it x, can only take on a discrete set of values $\{x_1, x_2, ..., x_M\}$ then we can define the probability distribution

$$P(x_i) = \text{the probability that a single measurement}$$
$$\text{will give a value } x_i \in \{x_1, x_2, ..., x_M\}. \tag{B.8}$$

If the random variable can take on any value in a continuous range then we define the probability distribution

$$P(x)\, dx = \text{the probability that a single measurement}$$
$$\text{will give a value between } x \text{ and } x + dx. \tag{B.9}$$

By relating statistics like the mean and the standard deviation to the underlying probability distributions we can relate probabilities to our estimated uncertainties. We typically only know the results of our observations and don't know the underlying probability distribution, but as we'll learn in the next sections we usually have a good idea of what the probability distribution should be. To see how to relate probability distributions to statistics lets start with the definition of the mean in equation (B.4) and a set of N data points $\{y_1, y_2, ..., y_N\}$. For a discrete distribution this can be rewritten in terms of the discrete set of possible random variable values $\{x_1, x_2, ..., x_M\}$,

$$\bar{y} = \frac{1}{N} \sum_{i=1}^{N} y_i = \frac{1}{N} \sum_{j=1}^{M} n_j x_j = \sum_{j=1}^{M} \frac{n_j}{N} x_j,$$

where n_j is the number of times the discrete random variable value x_j appears in the data set. Notice the subtle distinction between the random variable x and the set of data points $\{y_1, y_2, ..., y_N\}$; all of the x values are unique[2]. The probability of finding any one of the random variable values x_j in the data set is

$$P(x_j) = \frac{n_j}{N},$$

therefore

$$\bar{y} = \sum_{j=1}^{M} P(x_j) x_j.$$

I have used x and y to make the distinction between the the random variable and the actual data values obvious, but the usual convention is to let $\bar{x}$ represent the mean,

$$\bar{x} = \sum_{j=1}^{M} x_j P(x_j). \tag{B.10}$$

For a continuous distribution the sum in equation (B.10) becomes an integration over the range of the random variable

$$\bar{x} = \int_a^b x P(x) dx, \tag{B.11}$$

where the range of x is between a and b. You can compute any statistic from the probability distribution. For example, the population standard deviation is

$$s_p^2 = \sum_{j=1}^{M} (x_j - \bar{x})^2 P(x_j), \tag{B.12}$$

for a discrete distribution or

$$s_p^2 = \int_a^b (x - \bar{x})^2 P(x) dx \tag{B.13}$$

for a continuous distribution.

---

[2] Here's an example to clarify the distinction. Suppose we have a set of five data points $y_1 = 4$, $y_2 = 6$, $y_3 = 8$, $y_4 = 6$, $y_5 = 8$. There are only three distinct values of the random variable $x_1 = 4$, $x_2 = 6$, $x_3 = 8$. For this data set $n_1 = 1$, $n_2 = 2$, and $n_3 = 2$.

B.6.1   The Normal Distribution

The normal, or Gaussian, distribution is the most common distribution for most random variables encountered in physics and astronomy. In Section B.6.3, we will learn that even in cases where the random variable isn't governed by a normal distribution, the means derived from that distribution are normally distributed. The normal distribution is a continuous distribution defined by

$$P_n(x, \mu, \sigma)\, dx = \frac{1}{\sigma\sqrt{2\pi}} e^{-\frac{1}{2}\left(\frac{x-\mu}{\sigma}\right)^2} dx, \qquad (B.14)$$

and is shown in Figure B.3. A random variable governed by the Gaussian

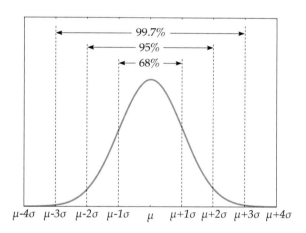

Figure B.3: The normal distribution. The distribution is centered on μ and has a width defined by σ. A number drawn from this distribution has a 68% chance of having a value within 1σ of μ, a 95% chance of being within 2σ of μ, and a 99.7% chance being within 3σ of μ.

distribution is said to be normally distributed.

You can use equation (B.11) to show that the mean of the distribution is μ and equation (B.13) to show that the population standard deviation is σ. When we make a measurement of some physical quantity that is governed by a normal distribution we have 68% chance of getting a value between $\mu - \sigma$ and $\mu + \sigma$, a 95% chance of getting a value between $\mu - 2\sigma$ and $\mu + 2\sigma$, and a 98.7% chance of getting a value between

$\mu - 3\sigma$ and $\mu + 3\sigma$. If we take the mean of several measurements we get an estimate of μ,

$$\bar{x} \approx \mu. \tag{B.15}$$

The *sample* standard deviation gives an estimate of σ,

$$s \approx \sigma. \tag{B.16}$$

See Bevington and Robinson [3] for the proofs of these properties.

Problem B.4
Numerically integrate the normal distribution from $\mu - \sigma$ to $\mu + \sigma$ to verify that the probability of getting x in the range $\mu - \sigma < x < \mu + \sigma$ is 0.68.

If we measure a quantity governed by a normal distribution then the best estimate of the value of the quantity is $\bar{x}$ and the uncertainty in a *single* measurement is related to σ. You can usually assume that the uncertainties quoted in the literature are so called "one-sigma" uncertainties, that is

$$\delta x = s \approx \sigma. \tag{B.17}$$

This means that there is a 68% chance that a single measurement is within $1\,\sigma$ of μ. If you measured a quantity by computing the mean then the appropriate uncertainty is the *standard deviation of the mean*,

$$\delta x = \frac{s}{\sqrt{N}} \approx \sigma_{\text{mean}}, \tag{B.18}$$

where N is the number of data points used to compute the mean. This means that if you repeated the series of N measurements, 68% of the time you would get a value within $1\,\sigma_{\text{mean}}$ of μ.

B.6.2 The Poisson Distribution

Even with a perfect, noise-free detector we cannot measure the brightness of a star with absolute certainty. This is because the emission and detection of a photon is probabilistic. If we used a noiseless detector to count the number of photons from a star in a given period of time, there would be no uncertainty in number of photons registered by the detector. However, if we repeated the measurement we would get a different number of photons because the probability of detection of a photon is

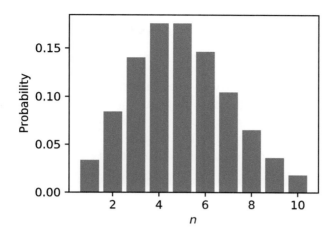

Figure B.4: The Poisson distribution with an mean of 5. Note that the standard deviation of the distribution is $\sqrt{5}$.

governed by quantum mechanics. The probability distribution for detection is called the **Poisson distribution**. It is a discrete probability distribution because we will always detect a whole number of photons. The probability that we will count n photons in a given trial is

$$P_p(n, \nu) = \frac{\nu^n}{n!}e^{-\nu}, \tag{B.19}$$

where ν is related to the probability of detection. The Poisson distribution for $\nu = 5$ is shown in Figure B.4. The Poisson distribution is the underlying probability distribution for all counting experiments where the measurements are uncorrelated and occur at a constant average rate.

Problem B.5

Use equation (B.10) to prove that the mean of the Poisson distribution is ν.

The mean of the Poisson distribution is ν, but the Poisson distribution also has the the unique property that the population standard deviation of the distribution, s_p, is equal to the square root of its mean,

$$s_p = \sqrt{\nu}. \tag{B.20}$$

Problem B.6
Use equation (B.12) to prove that $s_p = \sqrt{\nu}$ for the Poisson distribution.

This gives us a convenient way to estimate the both the photon count rate and the uncertainty in the rate from a single measurement. For a single measurement of n counts, n is a reasonable estimate of ν. An estimate of the uncertainty in n is therefore $\delta n = s_p \approx \sqrt{n}$. Our best estimate of the number of counts with uncertainty is therefore

$$n \pm \sqrt{n}. \tag{B.21}$$

The **fractional uncertainty** is defined as

$$\frac{\delta n}{n_{\text{best}}}. \tag{B.22}$$

For a counting experiment

$$\frac{\delta n}{n_{\text{best}}} = \frac{\sqrt{n}}{n} = \frac{1}{\sqrt{n}}. \tag{B.23}$$

The fractional uncertainty goes down as the number of counts increases. Counting more photons produces a smaller fractional uncertainty. One way to increase the number of photons detected from a star would be to increase the exposure time. The number of counts is proportional to the exposure time, t, so the fractional uncertainty is

$$\frac{\delta n}{n_{\text{best}}} \propto \frac{1}{\sqrt{t}}.$$

In order to reduce the fractional uncertainty by a factor of two we would have to increase the exposure time by a factor of four.

Problem B.7
An astronomer is preparing to measure the brightness of a star. He expects the photon count rate in his detector to be about 100 photons/second. Assuming the only source of noise in the measurement comes from photon counting, what exposure time would the astronomer need to use to get a measurement accurate to 1%?

B.6.3 The Central Limit Theorem

There are a huge number of different possible probability distributions, but fortunately we usually only encounter these two. Most of the time the random variables we measure are distributed normally. This is because most measurements are averages of individual observations and the **Central Limit Theorem** states that the *averages* of a large number of measurements of a random variable are normally distributed regardless of the underlying distribution of the random variable.

Here is an example to see how this works. Suppose you go to a daycare center and measure the heights of all the children and parents in the room. If it's a large daycare with about 500 children and 500 parents, you might get a distribution that looks like Figure B.5(a). The heights are not normally distributed. The children's heights are normally distributed about a mean of about 140 cm and the adults around a mean of 180 cm. The combined distribution of children and adults is *not* normally distributed. Now you randomly sample five people from the room and compute the mean height of the group. Your five person groups will have a random mix of children and adults. The mean height of the group will most likely be between 140 cm and 180 cm. After doing this several times you would get a distribution like the on shown in Figure B.5(b). It is beginning to look more like a normal distribution. If you repeated the process with ten people in each sample, the distribution would look even more like a normal distribution (Figure B.5(c)). The individual heights are *not* normally distributed, but the mean heights are closer to a normal distribution. The larger the number of samples in the mean the closer the distribution is to being a normal distribution. In addition to the distribution becoming more Gaussian, the width of the distribution decreases as more heights are included in the mean. It turns out that the width of the distribution of means is proportional to $1/\sqrt{N}$ where N is the number of values used to compute the mean. See Bevington and Robinson [3] for a derivation of the central limit theorem and a discussion of its limitations.

Even the Poisson distribution approaches the normal distribution for a large number of counts. Figure B.6 shows a Poisson distribution with $\nu = 20$ and a normal distribution with $\mu = 20$ and $\sigma = \sqrt{20}$. The normal distribution is a surprisingly good approximation to the Poisson distribution for even a relatively small number of counts.

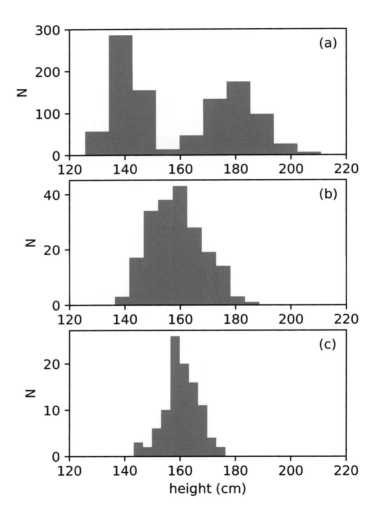

Figure B.5: Histogram (a) shows the distribution of 1000 heights of parents and children in a daycare center. The histogram (b) is the distribution of mean heights of five people chosen at random from the parent-child distribution. Histogram (c) is the distribution of mean heights of ten people chosen at random from the parent-child distribution.

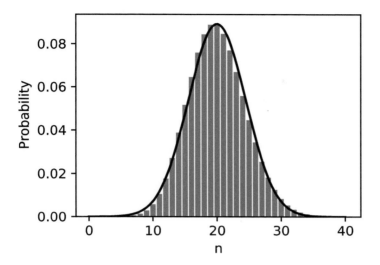

Figure B.6: Comparison of a Poisson distribution with $\nu = 20$ (histogram) and normal distribution (smooth curve) for $\mu = 20$ and $\sigma = \sqrt{20}$.

B.7 PROPAGATION OF UNCERTAINTY

Suppose we wanted to determine the perimeter and area of a rectangular table top from measurements of its width $w = 76.4 \pm 0.2$ cm and its length $\ell = 153.3 \pm 0.6$ cm. How would we determine the uncertainties in the perimeter and area? Let's compute the perimeter first. The perimeter $p = 2w + 2\ell = 2(w + \ell)$. The highest probable value for the perimeter is then

$$p_{\max} = 2(w + \delta w) + 2(\ell + \delta \ell) = 2(w + \ell) + 2(\delta w + \delta \ell),$$

and lowest probable value is

$$p_{\min} = 2(w - \delta w) + 2(\ell - \delta \ell) = 2(w + \ell) - 2(\delta w + \delta \ell).$$

We can use equation (B.3) to get the uncertainty in p

$$\delta p \approx \frac{p_{\max} - p_{\min}}{2} = 2(\delta w + \delta \ell).$$

For the example at hand, $\delta p = 1.6$ cm, so using the rounding rules of Section B.2 gives $p = 459 \pm 2$ cm.

You can always get a crude estimate of the uncertainty using this brute-force method of computing the maximum and minimum values and using equation (B.3), but it overestimates random uncertainties. This is, because to get the maximum uncertainty in the sum both ℓ and w must conspire to be overestimates. If the uncertainties are independent and random, then an underestimate in the measurement one of the variable, ℓ for example, is partially compensated for by an overestimate in the measurement of the other variable, w in this case. Statistical theory tells us that if some quantity $q = x + y$ and if x and y are normally distributed random variables, then the sum is also normally distributed with standard deviation

$$\sigma_q = \sqrt{\sigma_x^2 + \sigma_y^2}, \tag{B.24}$$

where σ_x and σ_y are the standard deviations of the x and y distributions (Bevington and Robinson [3]). This is always less than the sum of the standard deviations of x and y. When we combine two numbers by squaring them, adding the squares, and taking the square root as in equation (B.24), the numbers are said to be **added in quadrature.** Equation (B.24) would be the same if $q = x - y$ so in general we have the following rule:

If q is the sum or difference of several quantities $x, y, z...$

$$q = x + y - z \cdots$$

and if the uncertainties, $\delta x, \delta y, \delta z...$, are independent and random, then the uncertainty in q is the quadrature sum,

$$\delta q = \sqrt{\delta x^2 + \delta y^2 + \delta z + \cdots}. \tag{B.25}$$

There is a similar rule for products and quotients.

If several quantities $w, x, y, z...$ are measured with independent and random uncertainties $\delta w, \delta x, \delta y, \delta z...$ and

$$q = \frac{w \times x \times \cdots}{y \times z \times \cdots}$$

then the *fractional* uncertainty in q is the quadrature sum of the *factional* uncertainties in $w, x, y, z...,$

$$\frac{\delta q}{|q|} = \sqrt{\left(\frac{\delta w}{w}\right)^2 + \left(\frac{\delta x}{x}\right)^2 + \left(\frac{\delta y}{y}\right)^2 + \left(\frac{\delta z}{z}\right)^2 + \cdots}. \tag{B.26}$$

The area of the rectangle $A = w\ell = 11712.12 \text{ cm}^2$ and the fractional uncertainty in A is

$$\frac{\delta A}{|A|} = \sqrt{\left(\frac{\delta w}{w}\right)^2 + \left(\frac{\delta \ell}{\ell}\right)^2} = \sqrt{\left(\frac{0.2}{76.4}\right)^2 + \left(\frac{0.6}{153.3}\right)^2} = 0.0047,$$

so $\delta A = 55 \text{ cm}^2$. Using the rounding rules of Section B.2 gives $A = 11710 \pm 60 \text{ cm}^2$.

Sometimes the value we are interested in can't be written as a simple sum, difference, product, or quotient of the measured variables. In this case the uncertainties add in quadrature, but each uncertainty is weighted by a partial derivative.

> Suppose that some physical quantity $q(x, y, z, ...)$ is a function of measured values x, y, z,.... If the uncertainties, $\delta x, \delta y, \delta z...$, are independent and random, then the uncertainty in q is

$$\delta q = \sqrt{\left(\frac{\partial q}{\partial x}\delta x\right)^2 + \left(\frac{\partial q}{\partial y}\delta y\right)^2 + \left(\frac{\partial q}{\partial z}\delta z\right)^2}. \qquad (\text{B.27})$$

Problem B.8
Use equation (B.27) to derive equations (B.25) and (B.26).

When q is a complicated function of many variables, equation (B.27) can be formidable, but keep in mind that we only need to know δq to at most two significant figures. This means that if one of the terms under the radical in equation (B.27) is significantly larger than the others we can ignore all but the largest term.

SUPPLEMENTARY PROBLEMS

Problem B.9 Compute the uncertainty in the perimeter of the table discussed at the beginning of Section B.7 assuming the length and width measurements are independent. Hint: Equation (B.25) gives the wrong answer.

Problem B.10 The magnitude of a star is related to the photon count rate $\dot{n}$ by the equation

$$m = -2.5 \log(\dot{n}) + C,$$

where C is a constant. Show that the uncertainty in the magnitude

$$\delta m \approx \frac{\delta \dot{n}}{\dot{n}}.$$

Fitting and Graphical Representation of Data

C.1 "A PICTURE IS WORTH A THOUSAND WORDS"

Remember the old adage. A graph is the best way to make your data understandable when one observed quantity is depends on another. For example, suppose you measure the distance a car has traveled as time goes on and get the values shown shown in Table C.1.

Table C.1: Data for distance versus time measurement.

Time (seconds)	Distance (meters)
0.24	5.76×10^{-2}
1.29	1.68×10^{0}
2.35	5.55×10^{0}
3.41	1.16×10^{1}
4.47	2.00×10^{1}
5.53	3.06×10^{1}
6.59	4.34×10^{1}
7.65	5.85×10^{1}

It is very difficult to see how the distance is changing with time just by looking at the numbers in the table. Figure C.1 shows a graph of the data from Table C.1. It's easy to see from a quick look at the graph that the car is accelerating. It's very difficult to see this kind of trend by examining the data in a table. Take a close look at Figure C.1. Notice that the figure has clearly labeled axes including units. You don't need

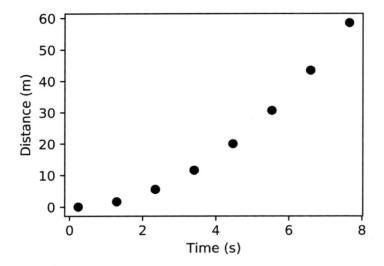

Figure C.1: Graph of the data in Table C.1.

to draw lines from data point to data points, in fact this is misleading in some cases.

You can represent uncertainties in your data by putting error bars on your graphs. Table C.2 contains data from a radioactive decay experiment where $\dot{N}$ is the decay rate as a function of time t. There is an estimated uncertainty in the rate. Figure C.2 shows a plot of these data with error bars to represent the uncertainty in $\dot{N}$.

Problem C.11

Type Ia supernovae are so luminous that they can be seen at very large distances. Hence, they are often used to determine the history of the expansion rate of the universe. The data file on the companion website called SN_gold.csv contains the redshifts and distances for some of the best observed supernova as of 2004.

(a) Download the file from the companion website and use a computer to plot the distance versus redshift. Don't forget to label the axes.

(b) Hubble measured nearby galaxies and proposed a linear relation-

---

Table C.2: Rate of radioactive decay $\dot{N}$ as a function of time t.

t (seconds)	$\dot{N}$
0.00	16 ± 4
0.05	15 ± 3
0.10	10 ± 3
0.15	12 ± 3
0.20	9 ± 3
0.25	7 ± 2
0.30	3 ± 1
0.35	5 ± 2
0.40	4 ± 2
0.45	1 ± 1

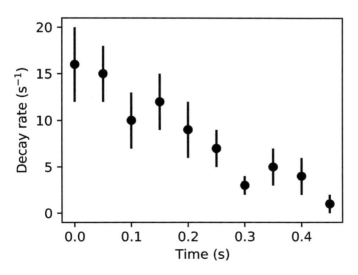

Figure C.2: Graph of the data in Table C.2.

> ship between redshift and distance. Do you see a linear relationship in your plot?

C.2 DATA FITTING

One of the most common data analysis problems is fitting a function to a set of data points. Suppose you have a set of N data points $\{x_i, y_i\}$ and you wish to fit the data to the function $y(x) = y(x; a_1, a_2, \cdots a_M)$, where the $\{a_j\}$ are adjustable parameters that give the "best fit." How do we define the "best fit"? We have many choices, but one of the most intuitive is to minimize the sum of the squares of the difference between the data points and the function. In other words, we wish to find the values of parameters $\{a_i\}$ that minimizes the function

$$S(a_1, a_2, \cdots a_M) = \sum_{i=1}^{M} [y_i - y(x_i; a_1, a_2, \cdots a_M)]^2 \qquad \text{(C.1)}$$

We can find the values of the $\{a_j\}$ that minimizes $S(a_1, a_2, \cdots a_M)$ by taking the partial derivatives of $S(a_1, a_2, \cdots a_M)$ with respect to the $\{a_j\}$ and setting the derivatives equal to zero. This will give a system M equations which we can solve to obtain the values of the $\{a_j\}$ that give extremal values of $S(a_1, a_2, \cdots a_M)$. The values of $\{a_j\}$ that give the absolute minimum of $S(a_1, a_2, \cdots a_M)$ are our best fit values. If the function is simple we can do the calculation analytically, but if the function is at all complicated we must resort to some computational method to solve the system. Fortunately, most numerical software packages include commands for doing curve fitting. This book's companion website describes how to use Python to fit data to most functions that you are likely to encounter.

Finding the minimum of works well $S(a_1, a_2, \cdots a_M)$ if the uncertainties in the individual data points are all the same. It is your only option if you don't know the uncertainties. Sometimes, however, you will have estimates of the uncertainties in each of the data points. If some of the data points have small uncertainties and others have large uncertainties, the points with large uncertainties can bias the fit. Figure C.3 shows linearly related data, but there is one point that has a huge uncertainty. This point could bias the fit if we used the technique above. We need a way to weigh the data points by their uncertainty so that points with a small error will count more than those with a large error. The

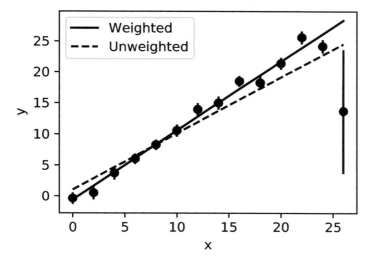

Figure C.3: Graph of linearly related data with one point that has a large uncertainty. The unweighted fit shows how the low data point with large uncertainty biases the fit. The weighted fit gives a better result.

standard way to approach such problems is to use maximum likelihood estimation (MLE) methods. The theory of MLE is beyond the scope of this book, but we can use the results. MLE theory tells us that if we have a data set where the errors in the y values are normally distributed, then instead of minimizing the function $S(a_1, a_2, \cdots a_M)$ we should minimize

$$\chi^2(a_1, a_2, \cdots a_M) = \sum_{i=1}^{N} \frac{[y_i - y(x_i; a_1, a_2, \cdots a_M)]^2}{\sigma_i^2}, \qquad (C.2)$$

where $\{\sigma_i\}$ are the uncertainties in the y data points. The solid line in Figure C.3 shows the fit generated by using a weighted fit. The weighted fit clearly gives a better result.

C.3   LINEAR FITTING

If the function you are trying to fit is linear, you can derive a closed form solution for the fit parameters and their uncertainties. In this section we outline the derivation of these equations and explore the interpretation of the $\chi^2(a_1, a_2, \cdots a_M)$ in more detail.

Suppose you have a set of N data pairs $\{x_i, y_i\}$, that are supposed to be linearly related so that

$$y = A + Bx. \tag{C.3}$$

The goal of a linear fit is to find the best fit parameters A and B. For a linear MLE fit, the equations for A and B are derived by finding the values of A and B that minimizes

$$\chi^2(A, B) = \sum_{i=1}^{N} \frac{(y_i - A - Bx_i)^2}{\sigma_i^2}. \tag{C.4}$$

The $\{\sigma_i\}$ values are the uncertainties in the y data values and we assume that the uncertainties in the x data values is negligible.

We can use the value of χ^2 to judge the quality of the fit. A fit is "good" if $\chi^2 \approx N - 2$. If $\chi^2 \ll N - 2$ it indicates that the uncertainties $\{\sigma_i\}$ have been overestimated. If $\chi^2 \gg N - 2$ then either the uncertainties have been underestimated or y isn't really linearly related to x by equation (C.3). See Bevington and Robinson [3] for a rigorous discussion of χ^2 and its interpretation.

Given the equations for A and B, one can use standard error propagation to derive general equations for their uncertainties. However, the values one accepts for these uncertainties depends on the quality of the fit. In the following sections, I outline how to estimate the uncertainties in A and B given something other than a perfect fit. The next section deals with the case in which the uncertainties in the y data points is all the same. The final section generalizes the technique to a weighted fit where the uncertainties for each data point are different.

C.3.1 Uncertainties for an Unweighted Linear Fit

In this section we assume that the y-uncertainty is the same for all the data. If we call this uncertainty σ, then

$$\chi^2 = \frac{1}{\sigma^2} \sum_{i=1}^{N} (y_i - A - Bx_i)^2. \tag{C.5}$$

The best estimates of A and B are those that minimize χ^2. Taking the partial derivatives of χ^2 with respect to A and B and setting them equal to zero gives

$$A = \frac{(\sum x_i^2)(\sum y_i) - (\sum x_i)(\sum x_i y_i)}{\Delta}, \qquad (C.6)$$

$$B = \frac{N(\sum x_i y_i) - (\sum x_i)(\sum y_i)}{\Delta}, \qquad (C.7)$$

where

$$\Delta = N(\sum x_i^2) - (\sum x_i)^2. \qquad (C.8)$$

We already know the uncertainty in the $\{y_i\}$ is σ, but can also use our fit to estimate what the uncertainties should be by looking at how much our data points deviate from the fit. The standard deviation from the fit gives us an estimated uncertainty

$$\sigma_{\text{est}} = \sqrt{\frac{1}{N-2} \sum_{i=1}^{N} (y_i - A - B x_i)^2}. \qquad (C.9)$$

By comparing this result with the equation (C.5) for χ^2 it is relatively easy to show that

$$\sigma_{\text{est}}^2 = \frac{\chi^2}{N-2} \sigma^2. \qquad (C.10)$$

At this point it is convenient to define the reduced chi-squared,

$$\tilde{\chi}^2 \equiv \frac{\chi^2}{N-2}. \qquad (C.11)$$

Given this definition, $\sigma_{\text{est}}^2 = \tilde{\chi}^2 \sigma^2$. Note that if we have a "good" fit, $\tilde{\chi}^2 = 1$ and $\sigma_{\text{est}} = \sigma$ as we would expect. In fact, this is what we mean by a good fit. If $\tilde{\chi}^2 \ll 1$ then $\sigma_{\text{est}} \ll \sigma$ and we have overestimated the error σ. If $\tilde{\chi}^2 \gg 1$ then $\sigma_{\text{est}} \gg \sigma$ and we have underestimated the error or the relation between x and y isn't really linear.

We can now use standard error propagation to find the uncertainties in A and B. It isn't too hard to show that

$$\sigma_A = \sqrt{\frac{\sum x_i^2}{\Delta}} \delta y,$$

$$\sigma_B = \sqrt{\frac{N}{\Delta}} \delta y,$$

where δy is the y-uncertainty. We now have a choice: do we let δy be our original uncertainty estimate σ or the uncertainty derived from the fit,

σ_{est}. If we don't have an uncertainty estimate σ then our only choice is to use σ_{est}. The estimated uncertainties in A and B are

$$\sigma_A = \sigma_{\text{est}} \sqrt{\frac{\sum x_i^2}{\Delta}}, \tag{C.12}$$

$$\sigma_B = \sigma_{\text{est}} \sqrt{\frac{N}{\Delta}}. \tag{C.13}$$

If we do know σ and $\tilde{\chi}^2 = 1$ the choice is moot since $\sigma = \sigma_{\text{est}}$. If the relation between x and y is truly linear, σ_{est} is best estimate of the error δy. Therefore, the best estimates for σ_A and σ_B are

$$\sigma_A = \sigma_{\text{est}} \sqrt{\frac{\sum x_i^2}{\Delta}} = \tilde{\chi}\sigma \sqrt{\frac{\sum x_i^2}{\Delta}}, \tag{C.14}$$

$$\sigma_B = \sigma_{\text{est}} \sqrt{\frac{N}{\Delta}} = \tilde{\chi}\sigma \sqrt{\frac{N}{\Delta}}. \tag{C.15}$$

However, be very cautious if you find $\sigma_{\text{est}} \gg \sigma$. Having a $\chi^2 > N - 2$ may mean the relation is not linear.

Problem C.12

Hubble observed several relatively nearby galaxies and concluded that their recessional velocity v is proportional to their distance d. Hubble actually measured the redshift of the galaxies. We can rewrite Hubble's law in terms of redshift z to get

$$d = \frac{c}{H_0} z,$$

were H_0 is the Hubble constant and we've assumed that $v \ll c$.

(a) Download the file SN_gold.csv from the companion website. Fit all of the supernova data in the file with $z < 0.1$ to a straight line. Use the slope to determine the Hubble constant. Don't forget to propagate the uncertainty in the slope.

(b) Is the intercept value consistent with zero?

(c) Create a plot showing the data and the fit. Don't forget to label the axes.

C.3.2 Uncertainties in a Weighted Fit

If the y uncertainties are not all equal then we can't factor them out of the sum in the expression for χ^2 as we did in equation (C.5). Instead we must use the more general expression for χ^2 given in equation (C.4). When we take the partial derivatives of equation (C.4) and set them equal to zero to minimize χ^2 we get the following equations for A and B:

$$A = \frac{(\sum w_i x_i^2)(\sum w_i y_i) - (\sum w_i x_i)(\sum w_i x_i y_i)}{\Delta}, \qquad (C.16)$$

$$B = \frac{(\sum w_i)(\sum w_i x_i y_i) - (\sum w_i x_i)(\sum w_i y_i)}{\Delta}, \qquad (C.17)$$

where

$$\Delta = (\sum w_i)(\sum w_i x_i^2) - (\sum w_i x_i)^2, \qquad (C.18)$$

and $w_i = 1/\sigma_i^2$. Applying error propagation to the equations for A and B allows us to compute the uncertainties in A and B. We find that

$$\sigma_A = \sqrt{\frac{\sum w_i x_i^2}{\Delta}},$$

$$\sigma_B = \sqrt{\frac{\sum w_i}{\Delta}}.$$

However, to derive these equations we have assumed that the σ_i are accurate estimates of the uncertainty so that $\tilde{\chi}^2 = 1$. If this is not the case then the above equations will either underestimate or overestimate the uncertainties in A and B. If we are confident that the the relation between x and y is linear, then just as in the case in section C.3.1, better estimates of the uncertainties are

$$\sigma_A = \tilde{\chi}\sqrt{\frac{\sum w_i x_i^2}{\Delta}}, \qquad (C.19)$$

$$\sigma_B = \tilde{\chi}\sqrt{\frac{\sum w_i}{\Delta}}. \qquad (C.20)$$

However, be cautious when using these equations. If $\tilde{\chi}^2 \gg 1$, it could mean that the relation between x and y is not linear. It could also mean your uncertainty estimates aren't reliable in which case it may be better to use an unweighted fit. However, if you are confident that the relation between x and y is linear, and that at least the relative size of the uncertainties are correct, then equations (C.19) and (C.20) give you the best estimates of the uncertainties in A and B respectively.

SUPPLEMENTARY PROBLEMS

Problem C.13 The density of dark matter in a galaxy decreases as you move radially from the center of a galaxy. One model predicts that the density profile is of the form

$$\rho(r) = \frac{\rho_0}{(r/a)(1 + r/a)^2},$$

where r is the radial distance from the center of the galaxy. The other two parameters ρ_0 and a are chosen to fit the observed density profile.

Download the file `GalProfile.csv` from the companion website. It contains simulated data of a galaxy's density profile. Fit the equation above to the simulated data to find the best fit parameter ρ_0 and a. Don't forget to determine the uncertainties in the parameters. Plot the data and the fit. How do they compare?

Bibliography

[1] M. S. Bessell. Standard photometric systems. *Annu. Rev. Astron. Astrophys.*, 43:293–336, 2005.

[2] M. S. Bessell, F. Caselli, and B. Pletz. Model atmospheres broadband colors, bolometric corrections and temperature calibrations for o-m stars. *Astron. Astrophys.*, 333:231–250, 1998.

[3] P. R. Bevington and D. K. Robinson. *Data reduction and error analysis for the physical sciences; 3rd ed.* McGraw-Hill Higher Education. McGraw-Hill, New York, NY, 2003.

[4] M. R. Blanton and S. Roweis. K-corrections and Filter Transformations in the Ultraviolet, Optical, and Near-infrared. *Astronomical Journal*, 133:734, February 2007.

[5] Hale Bradt. *Astronomy Methods: A Physical Approach to Astronomical Observations.* Cambridge University Press, 2004.

[6] B. W. Carroll and D. A. Ostlie. *An Introduction to Modern Astrophysics.* Addison-Wesley, second edition, 2006.

[7] Fredrick R. Chromey. *To Measure the Sky: An Introduction to Observational Astronomy.* Cambridge University Press, second edition, 2016.

[8] A. N. Cox, editor. *Allen's Astrophysical Quantities.* Springer, fourth edition, 1999.

[9] P. Duffett-Smith and J. Zwart. *Practical Astronomy with Your Calculator or Spreadsheet.* Cambridge University Press, fourth edition, 2011.

[10] A. S. Fruchter and R. N. Hook. Drizzle: A method for the linear reconstruction of undersampled images. *Publications of the Astronomical Society of the Pacific*, 114(792):144–152, 2002.

[11] M. Fukugita, T. Ichikawa, J. E. Gunn, M. Doi, K. Shimasaku, and D. P. Schneider. The Sloan Digital Sky Survey Photometric System. *Astronomical Journal*, 111:1748, Apr 1996.

[12] Robert H. Hardie. *Photoelectric Reductions*, page 178. 1964.

[13] H. L. Johnson and W. W. Morgan. Fundamental stellar photometry for standards of spectral type on the revised system of the Yerkes spectral atlas. *Astrophysical Journal*, 117:313, May 1953.

[14] Arlo U. Landolt. *UBVRI* photometric standard stars around the sky at +50 deg declination. *The Astronomical Journal*, 146(5):131, oct 2013.

[15] P. Martinez and A. Klotz. *A Practical Guide ot CCD Astronomy*, volume 8 of *Practical Astronomy Handbooks*. Cambridge University Press, 1997.

[16] I. Momcheva and E. Tollerud. Software use in astronomy: an informal survey, 2015.

[17] D. A. Neamen. *Semiconductor Physics And Devices: Basic Principles*. McGraw-Hill, 4th edition, 2011.

[18] D. R. Silva and M. E. Cornell. A new library of stellar optical spectra. *Astrophysical Journal*, 81:865–881, August 1992.

[19] W. M. Smart. *Textbook on Spherical Astronomy*. Cambridge University Press, sixth edition, 1977.

[20] J. A. Smith, D. L. Tucker, S. Kent, M. W. Richmond, M. Fukugita, T. Ichikawa, S.-i. Ichikawa, A. M. Jorgensen, A. Uomoto, J. E. Gunn, M. Hamabe, M. Watanabe, A. Tolea, A. Henden, J. Annis, J. R. Pier, T. A. McKay, J. Brinkmann, B. Chen, J. Holtzman, K. Shimasaku, and D. G. York. The u'g'r'i'z' Standard-Star System. *Astronomical Journal*, 123:2121–2144, April 2002.

[21] Robert Tyson. *Principles of Adaptive Optics*. Optics and Optoelectronics. CRC Press, third edition, 2010.

[22] Nautical Almanac Office (U.S.). *The Astronomical Almanac for the Year 2014*. U.S. Government Printing Office, 2013.